THE PRINCE

AND THE PAUPER

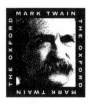

THE OXFORD MARK TWAIN

Shelley Fisher Fishkin, Editor

The Celebrated Jumping Frog of Calaveras County, and Other Sketches
Introduction: Roy Blount Jr.
Afterword: Richard Bucci

The Innocents Abroad
Introduction: Mordecai Richler
Afterword: David E. E. Sloane

Roughing It
Introduction: George Plimpton
Afterword: Henry B. Wonham

The Gilded Age
Introduction: Ward Just
Afterword: Gregg Camfield

Sketches, New and Old
Introduction: Lee Smith
Afterword: Sherwood Cummings

The Adventures of Tom Sawyer
Introduction: E. L. Doctorow
Afterword: Albert E. Stone

A Tramp Abroad
Introduction: Russell Banks
Afterword: James S. Leonard

The Prince and the Pauper
Introduction: Judith Martin
Afterword: Everett Emerson

Life on the Mississippi
Introduction: Willie Morris
Afterword: Lawrence Howe

Adventures of Huckleberry Finn
Introduction: Toni Morrison
Afterword: Victor A. Doyno

A Connecticut Yankee in King Arthur's Court
Introduction: Kurt Vonnegut, Jr.
Afterword: Louis J. Budd

Merry Tales
Introduction: Anne Bernays
Afterword: Forrest G. Robinson

The American Claimant
Introduction: Bobbie Ann Mason
Afterword: Peter Messent

The £1,000,000 Bank-Note and Other New Stories
Introduction: Malcolm Bradbury
Afterword: James D. Wilson

Tom Sawyer Abroad
Introduction: Nat Hentoff
Afterword: M. Thomas Inge

The Tragedy of Pudd'nhead Wilson and the Comedy
Those Extraordinary Twins
Introduction: Sherley Anne Williams
Afterword: David Lionel Smith

Personal Recollections of Joan of Arc
Introduction: Justin Kaplan
Afterword: Susan K. Harris

The Stolen White Elephant and Other Detective Stories
Introduction: Walter Mosley
Afterword: Lillian S. Robinson

How to Tell a Story and Other Essays
Introduction: David Bradley
Afterword: Pascal Covici, Jr.

Following the Equator and Anti-imperialist Essays
Introduction: Gore Vidal
Afterword: Fred Kaplan

The Man That Corrupted Hadleyburg and Other Stories and Essays
Introduction: Cynthia Ozick
Afterword: Jeffrey Rubin-Dorsky

The Diaries of Adam and Eve
Introduction: Ursula K. Le Guin
Afterword: Laura E. Skandera-Trombley

What Is Man?
Introduction: Charles Johnson
Afterword: Linda Wagner-Martin

The $30,000 Bequest and Other Stories
Introduction: Frederick Busch
Afterword: Judith Yaross Lee

Christian Science
Introduction: Garry Wills
Afterword: Hamlin Hill

Chapters from My Autobiography
Introduction: Arthur Miller
Afterword: Michael J. Kiskis

1601, and Is Shakespeare Dead?

Introduction: Erica Jong

Afterword: Leslie A. Fiedler

Extract from Captain Stormfield's Visit to Heaven

Introduction: Frederik Pohl

Afterword: James A. Miller

Speeches

Introduction: Hal Holbrook

Afterword: David Barrow

The Prince and the Pauper

Mark Twain

FOREWORD

SHELLEY FISHER FISHKIN

INTRODUCTION

JUDITH MARTIN

AFTERWORD

EVERETT EMERSON

New York Oxford

OXFORD UNIVERSITY PRESS

1996

OXFORD UNIVERSITY PRESS
Oxford New York
Athens, Auckland, Bangkok, Bogotá, Bombay
Buenos Aires, Calcutta, Cape Town, Dar es Salaam
Delhi, Florence, Hong Kong, Istanbul, Karachi
Kuala Lumpur, Madras, Madrid, Melbourne
Mexico City, Nairobi, Paris, Singapore
Taipei, Tokyo, Toronto
and associated companies in
Berlin, Ibadan

Text design by Richard Hendel
Composition: David Thorne

Published by
Oxford University Press, Inc.
198 Madison Avenue, New York,
New York 10016

Library of Congress
Cataloging-in-Publication Data

Twain, Mark, 1835–1910.
The prince and the pauper: a tale for young people
of all ages / by Mark Twain; with an introduction by
Judith Martin and an afterword by Everett Emerson.
p. cm. — (The Oxford Mark Twain)
Includes bibliographical references (p.).
1. Edward VI King of England, 1537–1553—Fiction.
2. Poor—England—Fiction. 3. Great Britain—Kings
and rulers—Fiction. I. Title. II. Series: Twain, Mark,
1835–1910. Works. 1996
PS1316.A1 1996
813'.4—dc20
96-17031
CIP
ISBN 0-19-510138-3 (trade ed.)
ISBN 0-19-511406-x (lib. ed.)
ISBN 0-19-509088-8 (trade ed. set)
ISBN 0-19-511345-4 (lib. ed. set)

9 8 7 6 5 4 3 2 1

Printed in the United States of America
on acid-free paper

FRONTISPIECE
The Clemens family posed for this photograph on the front porch of their Hartford home in 1884. Seated, from left to right, are Clara, Livy, Jean, Sam, and Susy. Clemens dedicated *The Prince and the Pauper* to his daughters Susy and Clara. (The Mark Twain House, Hartford, Connecticut)

CONTENTS

EDITOR'S NOTE

The Oxford Mark Twain consists of twenty-nine volumes of facsimiles of the first American editions of Mark Twain's works, with an editor's foreword, new introductions, afterwords, notes on the texts, and essays on the illustrations in volumes with artwork. The facsimiles have been reproduced from the originals unaltered, except that blank pages in the front and back of the books have been omitted, and any seriously damaged or missing pages have been replaced by pages from other first editions (as indicated in the notes on the texts).

In the foreword, introduction, afterword, and essays on the illustrations, the titles of Mark Twain's works have been capitalized according to modern conventions, as have the names of characters (except where otherwise indicated). In the case of discrepancies between the title of a short story, essay, or sketch as it appears in the original table of contents and as it appears on its own title page, the title page has been followed. The parenthetical numbers in the introduction, afterwords, and illustration essays are page references to the facsimiles.

FOREWORD

Shelley Fisher Fishkin

Samuel Clemens entered the world and left it with Halley's Comet, little dreaming that generations hence Halley's Comet would be less famous than Mark Twain. He has been called the American Cervantes, our Homer, our Tolstoy, our Shakespeare, our Rabelais. Ernest Hemingway maintained that "all modern American literature comes from one book by Mark Twain called *Huckleberry Finn*." President Franklin Delano Roosevelt got the phrase "New Deal" from *A Connecticut Yankee in King Arthur's Court*. *The Gilded Age* gave an entire era its name. "The future historian of America," wrote George Bernard Shaw to Samuel Clemens, "will find your works as indispensable to him as a French historian finds the political tracts of Voltaire."[1]

There is a Mark Twain Bank in St. Louis, a Mark Twain Diner in Jackson Heights, New York, a Mark Twain Smoke Shop in Lakeland, Florida. There are Mark Twain Elementary Schools in Albuquerque, Dayton, Seattle, and Sioux Falls. Mark Twain's image peers at us from advertisements for Bass Ale (his drink of choice was Scotch), for a gas company in Tennessee, a hotel in the nation's capital, a cemetery in California.

Ubiquitous though his name and image may be, Mark Twain is in no danger of becoming a petrified icon. On the contrary: Mark Twain lives. *Huckleberry Finn* is "the most taught novel, most taught long work, and most taught piece of American literature" in American schools from junior high to the graduate level.[2] Hundreds of Twain impersonators appear in theaters, trade shows, and shopping centers in every region of the country.[3] Scholars publish hundreds of articles as well as books about Twain every year, and he is the subject of daily exchanges on the Internet. A journalist somewhere in the world finds a reason to quote Twain just about every day. Television series such as *Bonanza*, *Star Trek: The Next Generation*, and *Cheers* broadcast episodes that feature Mark Twain as a character. Hollywood screenwriters regularly produce movies inspired by his works, and writers of mysteries and science fiction continue to weave him into their plots.[4]

A century after the American Revolution sent shock waves throughout Europe, it took Mark Twain to explain to Europeans and to his countrymen alike what that revolution had wrought. He probed the significance of this new land and its new citizens, and identified what it was in the Old World that America abolished and rejected. The founding fathers had thought through the political dimensions of making a new society; Mark Twain took on the challenge of interpreting the social and cultural life of the United States for those outside its borders as well as for those who were living the changes he discerned.

Americans may have constructed a new society in the eighteenth century, but they articulated what they had done in voices that were largely interchangeable with those of Englishmen until well into the nineteenth century. Mark Twain became the voice of the new land, the leading translator of what and who the "American" was — and, to a large extent, is. Frances Trollope's *Domestic Manners of the Americans,* a best-seller in England, Hector St. John de Crèvecoeur's *Letters from an American Farmer,* and Tocqueville's *Democracy in America* all tried to explain America to Europeans. But Twain did more than that: he allowed European readers to *experience* this strange "new world." And he gave his countrymen the tools to do two things they had not quite had the confidence to do before. He helped them stand before the cultural icons of the Old World unembarrassed, unashamed of America's lack of palaces and shrines, proud of its brash practicality and bold inventiveness, unafraid to reject European models of "civilization" as tainted or corrupt. And he also helped them recognize their own insularity, boorishness, arrogance, or ignorance, and laugh at it — the first step toward transcending it and becoming more "civilized," in the best European sense of the word.

Twain often strikes us as more a creature of our time than of his. He appreciated the importance and the complexity of mass tourism and public relations, fields that would come into their own in the twentieth century but were only fledgling enterprises in the nineteenth. He explored the liberating potential of humor and the dynamics of friendship, parenting, and marriage. He narrowed the gap between "popular" and "high" culture, and he meditated on the enigmas of personal and national identity. Indeed, it would be difficult to find an issue on the horizon today that Twain did not touch on somewhere in his work. Heredity versus environment? Animal rights? The boundaries of gender? The place of black voices in the cultural heritage of the United States? Twain was there.

With startling prescience and characteristic grace and wit, he zeroed in on many of the key challenges — political, social, and technological — that would face his country and the world for the next hundred years: the challenge of race relations in a society founded on both chattel slavery and ideals of equality, and the intractable problem of racism in American life; the potential of new technologies to transform our lives in ways that can be both exhilarating and terrifying — as well as unpredictable; the problem of imperialism and the difficulties entailed in getting rid of it. But he never lost sight of the most basic challenge of all: each man or woman's struggle for integrity in the face of the seductions of power, status, and material things.

Mark Twain's unerring sense of the right word and not its second cousin taught people to pay attention when he spoke, in person or in print. He said things that were smart and things that were wise, and he said them incomparably well. He defined the rhythms of our prose and the contours of our moral map. He saw our best and our worst, our extravagant promise and our stunning failures, our comic foibles and our tragic flaws. Throughout the world he is viewed as the most distinctively American of American authors — and as one of the most universal. He is assigned in classrooms in Naples, Riyadh, Belfast, and Beijing, and has been a major influence on twentieth-century writers from Argentina to Nigeria to Japan. The Oxford Mark Twain celebrates the versatility and vitality of this remarkable writer.

The Oxford Mark Twain reproduces the first American editions of Mark Twain's books published during his lifetime.[5] By encountering Twain's works in their original format — typography, layout, order of contents, and illustrations — readers today can come a few steps closer to the literary artifacts that entranced and excited readers when the books first appeared. Twain approved of and to a greater or lesser degree supervised the publication of all of this material.[6] The Mark Twain House in Hartford, Connecticut, generously loaned us its originals.[7] When more than one copy of a first American edition was available, Robert H. Hirst, general editor of the Mark Twain Project, in cooperation with Marianne Curling, curator of the Mark Twain House (and Jeffrey Kaimowitz, head of Rare Books for the Watkinson Library of Trinity College, Hartford, where the Mark Twain House collection is kept), guided our decision about which one to use.[8] As a set, the volumes also contain more than eighty essays commissioned especially for The Oxford Mark Twain, in which distinguished contributors reassess Twain's achievement as a writer and his place in the cultural conversation that he did so much to shape.

Each volume of The Oxford Mark Twain is introduced by a leading American, Canadian, or British writer who responds to Twain — often in a very personal way — as a fellow writer. Novelists, journalists, humorists, columnists, fabulists, poets, playwrights — these writers tell us what Twain taught them and what in his work continues to speak to them. Reading Twain's books, both famous and obscure, they reflect on the genesis of his art and the characteristics of his style, the themes he illuminated, and the aesthetic strategies he pioneered. Individually and collectively their contributions testify to the place Mark Twain holds in the hearts of readers of all kinds and temperaments.

Scholars whose work has shaped our view of Twain in the academy today have written afterwords to each volume, with suggestions for further reading. Their essays give us a sense of what was going on in Twain's life when he wrote the book at hand, and of how that book fits into his career. They explore how each book reflects and refracts contemporary events, and they show Twain responding to literary and social currents of the day, variously accepting, amplifying, modifying, and challenging prevailing paradigms. Sometimes they argue that works previously dismissed as quirky or eccentric departures actually address themes at the heart of Twain's work from the start. And as they bring new perspectives to Twain's composition strategies in familiar texts, several scholars see experiments in form where others saw only formlessness, method where prior critics saw only madness. In addition to elucidating the work's historical and cultural context, the afterwords provide an overview of responses to each book from its first appearance to the present.

Most of Mark Twain's books involved more than Mark Twain's words: unique illustrations. The parodic visual send-ups of "high culture" that Twain himself drew for *A Tramp Abroad*, the sketch of financial manipulator Jay Gould as a greedy and sadistic "Slave Driver" in *A Connecticut Yankee in King Arthur's Court*, and the memorable drawings of Eve in *Eve's Diary* all helped Twain's books to be sold, read, discussed, and preserved. In their essays for each volume that contains artwork, Beverly R. David and Ray Sapirstein highlight the significance of the sketches, engravings, and photographs in the first American editions of Mark Twain's works, and tell us what is known about the public response to them.

The Oxford Mark Twain invites us to read some relatively neglected works by Twain in the company of some of the most engaging literary figures of our time. Roy Blount Jr., for example, riffs in a deliciously Twain-like manner on "An Item Which the Editor Himself Could Not Understand," which may well rank as one of the least-known pieces Twain ever published. Bobbie Ann Mason celebrates the "mad energy" of Twain's most obscure comic novel, *The American Claimant*, in which the humor "hurtles beyond tall tale into simon-pure absurdity."[9] Garry Wills finds that *Christian Science* "gets us very close to the heart of American culture." Lee Smith reads "Political Economy" as a sharp and funny essay on language. Walter Mosley sees "The Stolen White Elephant," a story "reduced to a series of ridiculous telegrams related by an untrustworthy narrator caught up in an adventure that is as impossible as it is ludicrous," as a stunningly compact and economical satire of a world we still recognize as our own. Anne Bernays returns to "The Private History of a Campaign That Failed" and finds "an antiwar manifesto that is also confession, dramatic monologue, a plea for understanding and absolution, and a romp that gradually turns into atrocity even as we watch." After revisiting Captain Stormfield's heaven, Frederik Pohl finds that there "is no imaginable place more pleasant to spend eternity." Indeed, Pohl writes, "one would almost be willing to die to enter it."

While less familiar works receive fresh attention in The Oxford Mark Twain, new light is cast on the best-known works as well. Judith Martin ("Miss Manners") points out that it is by reading a court etiquette book that Twain's pauper learns how to behave as a proper prince. As important as etiquette may be in the palace, Martin notes, it is even more important in the slums.

> That etiquette is a sorer point with the ruffians in the street than with the proud dignitaries of the prince's court may surprise some readers. As in our own streets, etiquette is always a more volatile subject among those who cannot count on being treated with respect than among those who have the power to command deference.

And taking a fresh look at *Adventures of Huckleberry Finn,* Toni Morrison writes,

> much of the novel's genius lies in its quiescence, the silences that pervade it and give it a porous quality that is by turns brooding and soothing. It lies in . . . the subdued images in which the repetition of a simple word, such as "lonesome," tolls like an evening bell; the moments when nothing is said, when scenes and incidents swell the heart unbearably precisely because unarticulated, and force an act of imagination almost against the will.

Engaging Mark Twain as one writer to another, several contributors to The Oxford Mark Twain offer new insights into the processes by which his books came to be. Russell Banks, for example, reads *A Tramp Abroad* as "an important revision of Twain's incomplete first draft of *Huckleberry Finn*, a second draft, if you will, which in turn made possible the third and final draft." Erica Jong suggests that *1601,* a freewheeling parody of Elizabethan manners and mores, written during the same summer Twain began *Huckleberry Finn*, served as "a warm-up for his creative process" and "primed the pump for other sorts of freedom of expression." And Justin Kaplan suggests that "one of the transcendent figures standing behind and shaping" *Joan of Arc* was Ulysses S. Grant, whose memoirs Twain had recently published, and who, like Joan, had risen unpredictably "from humble and obscure origins" to become a "military genius" endowed with "the gift of command, a natural eloquence, and an equally natural reserve."

As a number of contributors note, Twain was a man ahead of his times. *The Gilded Age* was the first "Washington novel," Ward Just tells us, because "Twain was the first to see the possibilities that had eluded so many others." Commenting on *The Tragedy of Pudd'nhead Wilson,* Sherley Anne Williams observes that "Twain's argument about the power of environment in shaping character runs directly counter to prevailing sentiment where the negro was concerned." Twain's fictional technology, wildly fanciful by the standards of his day, predicts developments we take for granted in ours. DNA cloning, fax machines, and photocopiers are all prefigured, Bobbie Ann Mason tells us, in *The American Claimant.* Cynthia Ozick points out that the "telelectrophonoscope" we meet in "From the 'London Times' of 1904" is suspiciously like what we know as "television." And Malcolm Bradbury suggests that in the "phrenophones" of "Mental Telegraphy" "the Internet was born."

Twain turns out to have been remarkably prescient about political affairs as well. Kurt Vonnegut sees in *A Connecticut Yankee* a chilling foreshadowing (or perhaps a projection from the Civil War) of "all the high-tech atrocities which followed, and which follow still." Cynthia Ozick suggests that "The Man That Corrupted Hadleyburg," along with some of the other pieces collected under that title — many of them written when Twain lived in a Vienna ruled by Karl Lueger, a demagogue Adolf Hitler would later idolize — shoot up moral flares that shed an eerie light on the insidious corruption, prejudice, and hatred that reached bitter fruition under the Third Reich. And Twain's portrait in this book of "the dissolving Austria-Hungary of the 1890s," in Ozick's view, presages not only the Sarajevo that would erupt in 1914 but also "the disintegrated components of the former Yugoslavia" and "the *fin-de-siècle* Sarajevo of our own moment."

Despite their admiration for Twain's ambitious reach and scope, contributors to The Oxford Mark Twain also recognize his limitations. Mordecai Richler, for example, thinks that "the early pages of *Innocents Abroad* suffer from being a tad broad, proffering more burlesque than inspired satire," perhaps because Twain was "trying too hard for knee-slappers." Charles Johnson notes that the Young Man in Twain's philosophical dialogue about free will and determinism (*What Is Man?*) "caves in far too soon," failing to challenge what through late-twentieth-century eyes looks like "pseudoscience" and suspect essentialism in the Old Man's arguments.

Some contributors revisit their first encounters with Twain's works, recalling what surprised or intrigued them. When David Bradley came across "Fenimore Cooper's Literary Offences" in his college library, he "did not at first realize that Twain was being his usual ironic self with all this business about the 'nineteen rules governing literary art in the domain of romantic fiction,' but by the time I figured out there was no such list outside Twain's own head, I had decided that the rules made *sense*. . . . It seemed to me they were a pretty good blueprint for writing — Negro writing included." Sherley Anne Williams remembers that part of what attracted her to *Pudd'nhead Wilson* when she first read it thirty years ago was "that Twain, writing at the end of the nineteenth century, could imagine negroes as characters, albeit white ones, who actually thought for and of themselves, whose actions were the product of their thinking rather than the spontaneous ephemera of physical instincts that stereotype assigned to blacks." Frederik Pohl recalls his first reading of *Huckleberry Finn* as "a watershed event" in his life, the first book he read as a child in which "bad people" ceased to exercise a monopoly on doing "bad things." In *Huckleberry Finn* "some seriously bad things — things like the possession and mistreatment of black slaves, like stealing and lying, even like killing other people in duels — were quite often done by people who not only thought of themselves as exemplarily moral but, by any other standards I knew how to apply, actually *were* admirable citizens." The world that

Tom and Huck lived in, Pohl writes, "was filled with complexities and contradictions," and resembled "the world I appeared to be living in myself."

Other contributors explore their more recent encounters with Twain, explaining why they have revised their initial responses to his work. For Toni Morrison, parts of *Huckleberry Finn* that she "once took to be deliberate evasions, stumbles even, or a writer's impatience with his or her material," now strike her "as otherwise: as entrances, crevices, gaps, seductive invitations flashing the possibility of meaning. Unarticulated eddies that encourage diving into the novel's undertow — the real place where writer captures reader." One such "eddy" is the imprisonment of Jim on the Phelps farm. Instead of dismissing this portion of the book as authorial bungling, as she once did, Morrison now reads it as Twain's commentary on the 1880s, a period that "saw the collapse of civil rights for blacks," a time when "the nation, as well as Tom Sawyer, was deferring Jim's freedom in agonizing play." Morrison believes that Americans in the 1880s were attempting "to bury the combustible issues Twain raised in his novel," and that those who try to kick Huck Finn out of school in the 1990s are doing the same: "The cyclical attempts to remove the novel from classrooms extend Jim's captivity on into each generation of readers."

Although imitation-Hemingway and imitation-Faulkner writing contests draw hundreds of entries annually, no one has ever tried to mount a faux-Twain competition. Why? Perhaps because Mark Twain's voice is too much a part of who we are and how we speak even today. Roy Blount Jr. suggests that it is impossible, "at least for an American writer, to parody Mark Twain. It would be like doing an impression of your father or mother: he or she is already there in your voice."

Twain's style is examined and celebrated in The Oxford Mark Twain by fellow writers who themselves have struggled with the nuances of words, the structure of sentences, the subtleties of point of view, and the trickiness of opening lines. Bobbie Ann Mason observes, for example, that "Twain loved the sound of words and he knew how to string them by sound, like different shades of one color: 'The earl's barbaric eye,' 'the Usurping Earl,' 'a double-dyed humbug.'" Twain "relied on the punch of plain words" to show writers how to move beyond the "wordy romantic rubbish" so prevalent in nineteenth-century fiction, Mason says; he "was one of the first writers in America to deflower literary language." Lee Smith believes that "American writers have benefited as much from the way Mark Twain opened up the possibilities of first-person narration as we have from his use of vernacular language." (She feels that "the ghost of Mark Twain was hovering someplace in the background" when she decided to write her novel *Oral History* from the standpoint of multiple first-person narrators.) Frederick Busch maintains that "A Dog's Tale" "boasts one of the great opening sentences" of all time: "My father was a St. Bernard, my mother was a collie, but I am a Presbyterian." And Ursula Le Guin marvels at the ingenuity of the following sentence that she encounters in *Extracts from Adam's Diary*.

> . . . This made her sorry for the creatures which live in there, which she calls fish, for she continues to fasten names on to things that don't need them and don't come when they are called by them, which is a matter of no consequence to her, as she is such a numskull anyway; so she got a lot of them out and brought them in last night and put them in my bed to keep warm, but I have noticed them now and then all day, and I don't see that they are any happier there than they were before, only quieter.[10]

Le Guin responds,

> Now, that is a pure Mark-Twain-tour-de-force sentence, covering an immense amount of territory in an effortless, aimless ramble that seems to be heading nowhere in particular and ends up with breathtaking accuracy at the gold mine. Any sensible child would find that funny, perhaps not following all its divagations but delighted by the swing of it, by the word "numskull," by the idea of putting fish in the bed; and as that child grew older and reread it, its reward would only grow; and if that grown-up child had to write an essay on the piece and therefore earnestly studied and pored over this sentence, she would end up in unmitigated admiration of its vocabulary, syntax, pacing, sense, and rhythm, above all the beautiful

timing of the last two words; and she would, and she does, still find it funny.

The fish surface again in a passage that Gore Vidal calls to our attention, from *Following the Equator*: "'The Whites always mean well when they take human fish out of the ocean and try to make them dry and warm and happy and comfortable in a chicken coop,' which is how, through civilization, they did away with many of the original inhabitants. Lack of empathy is a principal theme in Twain's meditations on race and empire."

Indeed, empathy — and its lack — is a principal theme in virtually all of Twain's work, as contributors frequently note. Nat Hentoff quotes the following thoughts from Huck in *Tom Sawyer Abroad*:

> I see a bird setting on a dead limb of a high tree, singing with its head tilted back and its mouth open, and before I thought I fired, and his song stopped and he fell straight down from the limb, all limp like a rag, and I run and picked him up and he was dead, and his body was warm in my hand, and his head rolled about this way and that, like his neck was broke, and there was a little white skin over his eyes, and one little drop of blood on the side of his head; and laws! I could n't see nothing more for the tears; and I hain't never murdered no creature since that war n't doing me no harm, and I ain't going to.[11]

"The Humane Society," Hentoff writes, "has yet to say anything as powerful — and lasting."

Readers of The Oxford Mark Twain will have the pleasure of revisiting Twain's Mississippi landmarks alongside Willie Morris, whose own lower Mississippi Valley boyhood gives him a special sense of connection to Twain. Morris knows firsthand the mosquitoes described in *Life on the Mississippi* — so colossal that "two of them could whip a dog" and "four of them could hold a man down"; in Morris's own hometown they were so large during the flood season that "local wags said they wore wristwatches." Morris's Yazoo City and Twain's Hannibal shared a "rough-hewn democracy . . . complicated by all the visible textures of caste and class, . . . harmless boyhood fun and mischief right along with . . . rank hypocrisies, churchgoing sanctimonies, racial hatred, entrenched and unrepentant greed."

For the West of Mark Twain's *Roughing It*, readers will have George Plimpton as their guide. "What a group these newspapermen were!" Plimpton writes about Twain and his friends Dan De Quille and Joe Goodman in Virginia City, Nevada. "Their roisterous carryings-on bring to mind the kind of frat-house enthusiasm one associates with college humor magazines like the *Harvard Lampoon*." Malcolm Bradbury examines Twain as "a living example of what made the American so different from the European." And Hal Holbrook, who has interpreted Mark Twain on stage for some forty years, describes how Twain "played" during the civil rights movement, during the Vietnam War, during the Gulf War, and in Prague on the eve of the demise of Communism.

Why do we continue to read Mark Twain? What draws us to him? His wit? His compassion? His humor? His bravura? His humility? His understanding of who and what we are in those parts of our being that we rarely open to view? Our sense that he knows we can do better than we do? Our sense that he knows we can't? E. L. Doctorow tells us that children are attracted to *Tom Sawyer* because in this book "the young reader confirms his own hope that no matter how troubled his relations with his elders may be, beneath all their disapproval is their underlying love for him, constant and steadfast." Readers in general, Arthur Miller writes, value Twain's "insights into America's always uncertain moral life and its shifting but everlasting hypocrisies"; we appreciate the fact that he "is not using his alienation from the public illusions of his hour in order to reject the country implicitly as though he could live without it, but manifestly in order to correct it." Perhaps we keep reading Mark Twain because, in Miller's words, he "wrote much more like a father than a son. He doesn't seem to be sitting in class taunting the teacher but standing at the head of it challenging his students to acknowledge their own humanity, that is, their immemorial attraction to the untrue."

Mark Twain entered the public eye at a time when many of his countrymen considered "American culture" an oxymoron; he died four years before a world conflagration that would lead many to question whether the contradiction in terms was not "European civilization" instead. In between he worked in journalism, printing, steamboating, mining, lecturing, publishing, and editing, in virtually every region of the country. He tried his hand at humorous sketches, social satire, historical novels, children's books, poetry, drama, science fiction, mysteries, romance, philosophy, travelogue, memoir, polemic, and several genres no one had ever seen before or has ever seen since. He invented a self-pasting scrapbook, a history game, a vest strap, and a gizmo for keeping bed sheets tucked in; he invested in machines and processes designed to revolutionize typesetting and engraving, and in a food supplement called "Plasmon." Along the way he cheerfully impersonated himself and prior versions of himself for doting publics on five continents while playing out a charming rags-to-riches story followed by a devastating riches-to-rags story followed by yet another great American comeback. He had a long-running real-life engagement in a sumptuous comedy of manners, and then in a real-life tragedy not of his own design: during the last fourteen years of his life almost everyone he ever loved was taken from him by disease and death.

Mark Twain has indelibly shaped our views of who and what the United States is as a nation and of who and what we might become. He understood the nostalgia for a "simpler" past that increased as that past receded — and he saw through the nostalgia to a past that was just as complex as the present. He recognized better than we did ourselves our potential for greatness and our potential for disaster. His fictions brilliantly illuminated the world in which he lived, changing it — and us — in the process. He knew that our feet often danced to tunes that had somehow remained beyond our hearing; with perfect pitch he played them back to us.

My mother read *Tom Sawyer* to me as a bedtime story when I was eleven. I thought Huck and Tom could be a lot of fun, but I dismissed Becky Thatcher as a bore. When I was twelve I invested a nickel at a local garage sale in a book that contained short pieces by Mark Twain. That was where I met Twain's Eve. Now, *that's* more like it, I decided, pleased to meet a female character I could identify *with* instead of against. Eve had spunk. Even if she got a lot wrong, you had to give her credit for trying. "The Man That Corrupted

Hadleyburg" left me giddy with satisfaction: none of my adolescent reveries of getting even with my enemies were half as neat as the plot of the man who got back at that town. "How I Edited an Agricultural Paper" set me off in uncontrollable giggles.

People sometimes told me that I looked like Huck Finn. "It's the freckles," they'd explain — not explaining anything at all. I didn't read *Huckleberry Finn* until junior year in high school in my English class. It was the fall of 1965. I was living in a small town in Connecticut. I expected a sequel to *Tom Sawyer*. So when the teacher handed out the books and announced our assignment, my jaw dropped: "Write a paper on how Mark Twain used irony to attack racism in *Huckleberry Finn*."

The year before, the bodies of three young men who had gone to Mississippi to help blacks register to vote — James Chaney, Andrew Goodman, and Michael Schwerner — had been found in a shallow grave; a group of white segregationists (the county sheriff among them) had been arrested in connection with the murders. America's inner cities were simmering with pent-up rage that began to explode in the summer of 1965, when riots in Watts left thirty-four people dead. None of this made any sense to me. I was confused, angry, certain that there was something missing from the news stories I read each day: the why. Then I met Pap Finn. And the Phelpses.

Pap Finn, Huck tells us, "had been drunk over in town" and "was just all mud." He erupts into a drunken tirade about "a free nigger . . . from Ohio — a mulatter, most as white as a white man," with "the whitest shirt on you ever see, too, and the shiniest hat; and there ain't a man in town that's got as fine clothes as what he had."

> . . . they said he was a p'fessor in a college, and could talk all kinds of languages, and knowed everything. And that ain't the wust. They said he could *vote*, when he was at home. Well, that let me out. Thinks I, what is the country a-coming to? It was 'lection day, and I was just about to go and vote, myself, if I warn't too drunk to get there; but when they told me there was a State in this country where they'd let that nigger vote, I drawed out. I says I'll never vote agin. Them's the very words I said. . . . And to see the

> cool way of that nigger — why, he wouldn't a give me the road if I hadn't shoved him out o' the way.[12]

Later on in the novel, when the runaway slave Jim gives up his freedom to nurse a wounded Tom Sawyer, a white doctor testifies to the stunning altruism of his actions. The Phelpses and their neighbors, all fine, upstanding, well-meaning, churchgoing folk,

> agreed that Jim had acted very well, and was deserving to have some notice took of it, and reward. So every one of them promised, right out and hearty, that they wouldn't curse him no more.
>
> Then they come out and locked him up. I hoped they was going to say he could have one or two of the chains took off, because they was rotten heavy, or could have meat and greens with his bread and water, but they didn't think of it.[13]

Why did the behavior of these people tell me more about why Watts burned than anything I had read in the daily paper? And why did a drunk Pap Finn railing against a black college professor from Ohio whose vote was as good as his own tell me more about white anxiety over black political power than anything I had seen on the evening news?

Mark Twain knew that there was nothing, absolutely *nothing*, a black man could do — including selflessly sacrificing his freedom, the only thing of value he had — that would make white society see beyond the color of his skin. And Mark Twain knew that depicting racists with chilling accuracy would expose the viciousness of their world view like nothing else could. It was an insight echoed some eighty years after Mark Twain penned Pap Finn's rantings about the black professor, when Malcolm X famously asked, "Do you know what white racists call black Ph.D.'s?" and answered, "'*Nigger!*'"[14]

Mark Twain taught me things I needed to know. He taught me to understand the raw racism that lay behind what I saw on the evening news. He taught me that the most well-meaning people can be hurtful and myopic. He taught me to recognize the supreme irony of a country founded in freedom that continued to deny freedom to so many of its citizens. Every time I hear of another effort to kick Huck Finn out of school somewhere, I recall everything that Mark Twain taught *this* high school junior, and I find myself jumping into the fray.[15] I remember the black high school student who called CNN during the phone-in portion of a 1985 debate between Dr. John Wallace, a black educator spearheading efforts to ban the book, and myself. She accused Dr. Wallace of insulting her and all black high school students by suggesting they weren't smart enough to understand Mark Twain's irony. And I recall the black cameraman on the *CBS Morning News* who came up to me after he finished shooting another debate between Dr. Wallace and myself. He said he had never read the book by Mark Twain that we had been arguing about—but now he really wanted to. One thing that puzzled him, though, was why a white woman was defending it and a black man was attacking it, because as far as he could see from what we'd been saying, the book made whites look pretty bad.

As I came to understand *Huckleberry Finn* and *Pudd'nhead Wilson* as commentaries on the era now known as the nadir of American race relations, those books pointed me toward the world recorded in nineteenth-century black newspapers and periodicals and in fiction by Mark Twain's black contemporaries. My investigation of the role black voices and traditions played in shaping Mark Twain's art helped make me aware of their role in shaping all of American culture.[16] My research underlined for me the importance of changing the stories we tell about who we are to reflect the realities of what we've been.[17]

Ever since our encounter in high school English, Mark Twain has shown me the potential of American literature and American history to illuminate each other. Rarely have I found a contradiction or complexity we grapple with as a nation that Mark Twain had not puzzled over as well. He insisted on taking America seriously. And he insisted on *not* taking America seriously: "I think that there is but a single specialty with us, only one thing that can be called by the wide name 'American,'" he once wrote. "That is the national devotion to ice-water."[18]

Mark Twain threw back at us our dreams and our denial of those dreams, our greed, our goodness, our ambition, and our laziness, all rattling around together in that vast echo chamber of our talk — that sharp, spunky American talk that Mark Twain figured out how to write down without robbing it of its energy and immediacy. Talk shaped by voices that the official arbiters of "culture" deemed of no importance — voices of children, voices of slaves, voices of servants, voices of ordinary people. Mark Twain listened. And he made us listen. To the stories he told us, and to the truths they conveyed. He still has a lot to say that we need to hear.

Mark Twain lives — in our libraries, classrooms, homes, theaters, movie houses, streets, and most of all in our speech. His optimism energizes us, his despair sobers us, and his willingness to keep wrestling with the hilarious and horrendous complexities of it all keeps us coming back for more. As the twenty-first century approaches, may he continue to goad us, chasten us, delight us, berate us, and cause us to erupt in unrestrained laughter in unexpected places.

NOTES

1. Ernest Hemingway, *Green Hills of Africa* (New York: Charles Scribner's Sons, 1935), 22. George Bernard Shaw to Samuel L. Clemens, July 3, 1907, quoted in Albert Bigelow Paine, *Mark Twain: A Biography* (New York: Harper and Brothers, 1912), 3:1398.

2. Allen Carey-Webb, "Racism and *Huckleberry Finn*: Censorship, Dialogue and Change," *English Journal* 82, no. 7 (November 1993):22.

3. See Louis J. Budd, "Impersonators," in J. R. LeMaster and James D. Wilson, eds., *The Mark Twain Encyclopedia* (New York: Garland Publishing Company, 1993), 389–91.

4. See Shelley Fisher Fishkin, "Ripples and Reverberations," part 3 of *Lighting Out for the Territory: Reflections on Mark Twain and American Culture* (New York: Oxford University Press, 1996).

5. There are two exceptions. Twain published chapters from his autobiography in the *North American Review* in 1906 and 1907, but this material was not published in book form in Twain's lifetime; our volume reproduces the material as it appeared in the *North American Review*. The other exception is our final volume, *Mark Twain's Speeches*, which appeared two months after Twain's death in 1910.

An unauthorized handful of copies of *1601* was privately printed by an Alexander Gunn of Cleveland at the instigation of Twain's friend John Hay in 1880. The first American edition authorized by Mark Twain, however, was printed at the United States Military Academy at West Point in 1882; that is the edition reproduced here.

It should further be noted that four volumes — *The Stolen White Elephant and Other Detective Stories*, *Following the Equator and Anti-imperialist Essays*, *The Diaries of Adam and Eve*, and *1601, and Is Shakespeare Dead?* — bind together material originally published separately. In each case the first American edition of the material is the version that has been reproduced, always in its entirety. Because Twain constantly recycled and repackaged previously published works in his collections of short pieces, a certain amount of duplication is unavoidable. We have selected volumes with an eye toward keeping this duplication to a minimum.

Even the twenty-nine-volume Oxford Mark Twain has had to leave much out. No edition of Twain can ever claim to be "complete," for the man was too prolix, and the file drawers of both ephemera and as yet unpublished texts are deep.

6. With the possible exception of *Mark Twain's Speeches*. Some scholars suspect Twain knew about this book and may have helped shape it, although no hard evidence to that effect has yet surfaced. Twain's involvement in the production process varied greatly from book to book. For a fuller sense of authorial intention, scholars will continue to rely on the superb definitive editions of Twain's works produced by the Mark Twain Project at the University of California at Berkeley as they become available. Dense with annotation documenting textual emendation and related issues, these editions add immeasurably to our understanding of Mark Twain and the genesis of his works.

7. Except for a few titles that were not in its collection. The American Antiquarian Society in Worcester, Massachusetts, provided the first edition of *King Leopold's Soliloquy*; the Elmer Holmes Bobst Library of New York University furnished the 1906–7 volumes of the *North American Review* in which *Chapters from My Autobiography* first appeared; the Harry Ransom Humanities Research Center at the University of Texas at Austin made their copy of the West Point edition of *1601* available; and the Mark Twain Project provided the first edition of *Extract from Captain Stormfield's Visit to Heaven.*

8. The specific copy photographed for Oxford's facsimile edition is indicated in a note on the text at the end of each volume.

9. All quotations from contemporary writers in this essay are taken from their introductions to the volumes of The Oxford Mark Twain, and the quotations from Mark Twain's works are taken from the texts reproduced in The Oxford Mark Twain.

10. *The Diaries of Adam and Eve*, The Oxford Mark Twain [hereafter OMT] (New York: Oxford University Press, 1996), p. 33.

11. *Tom Sawyer Abroad*, OMT, p. 74.

12. *Adventures of Huckleberry Finn*, OMT, p. 49–50.

13. Ibid., p. 358.

14. Malcolm X, *The Autobiography of Malcolm X*, with the assistance of Alex Haley (New York: Grove Press, 1965), p. 284.

15. I do not mean to minimize the challenge of teaching this difficult novel, a challenge for which all teachers may not feel themselves prepared. Elsewhere I have developed some concrete strategies for approaching the book in the classroom, including teaching it in the context of the history of American race relations and alongside books by black writers. See Shelley Fisher Fishkin, "Teaching *Huckleberry Finn*," in James S. Leonard, ed., *Making Mark Twain Work in the Classroom* (Durham: Duke University Press, forthcoming). See also Shelley Fisher Fishkin, *Was Huck Black? Mark Twain and African-American Voices* (New York: Oxford University Press, 1993), pp. 106–8, and a curriculum kit in preparation at the Mark Twain House in Hartford, containing teaching suggestions from myself, David Bradley, Jocelyn Chadwick-Joshua, James Miller, and David E. E. Sloane.

16. See Fishkin, *Was Huck Black?* See also Fishkin, "Interrogating 'Whiteness,' Complicating 'Blackness': Remapping American Culture," in Henry Wonham, ed., *Criticism and the Color Line: Desegregating American Literary Studies* (New Brunswick: Rutgers UP, 1996, pp. 251–90 and in shortened form in *American Quarterly* 47, no. 3 (September 1995):428–66.

17. I explore the roots of my interest in Mark Twain and race at greater length in an essay entitled "Changing the Story," in Jeffrey Rubin-Dorsky and Shelley Fisher Fishkin, eds., *People of the Book: Thirty Scholars Reflect on Their Jewish Identity* (Madison: U of Wisconsin Press, 1996), pp. 47–63.

18. "What Paul Bourget Thinks of Us," *How to Tell a Story and Other Essays*, OMT, p. 197.

INTRODUCTION

Judith Martin

The sprig of Tudor royalty who was to become Edward VI is "struck dumb with amazement." How could he be impersonated successfully, in his very own court, by an urchin whom he had arbitrarily plucked from the streets only days before? "It did not seem possible that this could be, for surely his manners and speech would betray him if he pretended to be the Prince of Wales," the true prince reflects.

One can understand that this teenager would be stunned at the premise of Mark Twain's switcheroo plot. Having invested his entire childhood in learning and practicing the measured ways of court life, is he to believe that a raw recruit could be transformed into instant royalty, to the evident satisfaction of his own rigid and tireless instructors, and even his notoriously edgy father? "Why did I bother?" one can imagine him asking himself bitterly.

The device that makes this story possible is — of all things — an etiquette book. The look-alike pauper, Tom Canty, stranded in the clothes, persona and apartment of the prince, stumbles upon a book "about the etiquette of the English court. This was a prize. He lay down upon a sumptuous divan and proceeded to instruct himself with honest zeal." And arose from it a more or less properly behaved prince. (What he did about his speech patterns and accent, which would have been a lot harder to explain than behavioral lapses that are attributed to fatigue bordering on madness, is not known. Presumably, he also found language tapes to instruct him in the proper vocabulary and pronunciation.)

The book he read would most likely have been *The Babees Book* of 1475, which was specifically addressed to the princes of the English royal family, in an attempt to teach them the newly popular concept of restraining in public their natural inclinations, such as spitting, yawning, whispering and picking the nose. It could also have been Erasmus's *De Civilitate morum puerilium libellus*, published in 1526 and known at Henry VIII's court, although the philosopher aimed his similar instructions at less socially privileged students, with the radical plan of creating an intellectual elite that would be presentable (and therefore, he hoped, influential) in the highest circles of government.

Tom discovers this emergency etiquette assistance none too soon. He has already suffered agonies by not knowing whether he should scratch his own itchy nose or—having observed that princes have special dignitaries assigned to do even the humblest personal tasks for them—wait for a Hereditary Scratcher to do it for him.

Whichever etiquette book Tom consulted would have offered him instruction, but no relief. While the absence of mention of an official scratcher could be taken to be definitive, both books explicitly forbade boys to scratch themselves. Etiquette writers of the fifteenth and sixteenth centuries took a great interest in this point without, however, feeling obliged to offer alternative solutions.

The anti-scratching rule was part of a larger Renaissance response to the age-old philosophical question of what constitutes proper human behavior. The history of manners is a constant swing of the pendulum between the artificial and the natural—or, as these styles of behavior are termed by their respective detractors, the affected and the disgusting. During a time of elaborate conventions, people yearn for simplicity; during a time of frank crudeness, they yearn for refinement. Thus, Renaissance philosophers were promoting artificial improvements on the natural medieval manners to which people were sick of being exposed; Victorians reacted to what they considered the artificiality of the eighteenth century with the cult of what they called "sincerity"; the twentieth century reacted to what was seen as *Victorian* artificiality with the cult of "honesty"; and we are now beginning to have had enough of the twentieth century's idealized naturalness and are again receptive to the idea of a little unnatural politeness.

What makes *The Prince and the Pauper* a novel of manners, rather than merely a novel of cultural advantages, is its contribution to this debate, particularly to its most troubling aspect — that outward forms do not necessarily reflect inner character. The realizations that a virtuous heart does not ensure pleasant manners and that an evil heart can be disguised by them so dismay modern moralists that they can hardly bear to contemplate the very subject of manners, even though it was of such importance to their most distinguished predecessors.

Far from being stymied by that alarming paradox, this tale begins from the opposite assumption — that interior virtue is naturally linked to outward politeness. Although the author's preference for the scrappy plebeian over the stripling prince is clear, both boys are depicted as being of exceptionally good character, and at the book's opening, each demonstrates that his manners transcend the roughness of his particular background. Tom exhibits gentleness and dignity that are conspicuously lacking in his associates, and the prince is sensitive and hospitable in a way lacking among his. Thus (provided one generously interprets Tom's early habit of playing at being a prince as not marking him for a snob or a ninny, but indicating an inborn aptitude for gentility), both the prince and the pauper illustrate the idea that a good heart inspires a gracious manner.

Yet in the course of the book, they both have dangerous etiquette problems. In spite of — or perhaps because of — Tom's misgivings about his behavior, etiquette turns out to be much more of a problem for the prince, who tries to use his royal manners among common people, than for the commoner willingly acting as a prince, but both of them arouse suspicion and antagonism. This suggests that meaning well is not enough if one is not familiar with the etiquette specific to the circumstances in which one finds oneself. Good intentions fail to make up for misreading the social context.

In addition to their relationship to morals, manners are therefore a force in the plot, in that they are a device to place individuals within the society.

Mistaken identity depends on similarity, and when the author posits the mix-up of two boys from the extreme ends of the social scale, it is not enough to make them look alike. They must behave in unfamiliar worlds in such a way that neither is identified as what he really is. The pauper, when he decides he first needs, and then wants, to pass himself off as a prince, must practice manners that do not give him away as an impostor. It is equally necessary that the prince, who wants and demands to be recognized and treated as a prince, must practice manners that can be plausibly taken for those of a deluded pauper.

The setting takes for granted a world in which conditions and behavior vary widely but everyone is etiquette-conscious. Both spheres, palace and slum, are filled with people very much alive to transgressions of the standards they know and expect. While they have different notions of what constitutes offensive behavior, they share the concept that certain attitudes and conventions must be observed.

The poor, harboring a high sense of their own etiquette, are no less condemning of pretension and arrogance, and no less concerned with hierarchy and suitable demeanor, because their ideas on these matters are at variance with court procedure. They are quick to make charges of rudeness, and they are right, given their assumption that Edward is one of them, in thinking that he is badly behaved. What is majestic in an acknowledged prince is arrogant among equals. For a prince to condescend to a commoner is flattering; for one commoner to condescend to another is insulting.

The prince's inappropriate attitude to his presumed position, as well as his attempts to enforce a protocol that turns common practice — not to mention common sense — upside down by demanding that grown-ups show deference to a youngster, earn him life-threatening condemnation. While Tom keeps improving his new position by learning to behave like a prince, Edward keeps bringing disaster on himself by refusing to behave any other way. Even the ingratiating upper-class concept of *noblesse oblige*, which would require him to be especially polite to his inferiors, only applies to a rank just beneath his own. The consequences of etiquette transgression are also more severe in the lower and criminal classes than among courtiers — and not only because the former are given to blunter expression of disapproval while the latter have the habit of enforced tolerance and flexibility from dealing with those who outrank them and recognizing that royalty can make its own rules.

That etiquette is a sorer point with the ruffians in the street than with the proud dignitaries of the prince's court may surprise some readers. As in our own streets, etiquette is always a more volatile subject among those who cannot count on being treated with respect than among those who have the power to command deference. Far less significant than that Tom did yield and scratch his nose is the astonishing result: it passes unnoticed, because the assembled courtiers observe the highly sophisticated etiquette conventions of pretending that they failed to notice, and then of explaining away the supposedly unnoticed errors with the timeless excuses of illness and overwork.

This is not the response that Tom's peers would have made to an etiquette error (although they would not so define scratching an itch). They would have responded with ridicule or violence, and through long experience, Tom knows how to deal with both. Edward has to learn these techniques the hard way when, not knowing or caring that very different rules of etiquette prevail in his new surroundings, he deliberately or inadvertently violates them.

We see that at court it is considered rude to satirize people's (or maybe just princes') inadvertent transgressions. Whether that is praiseworthy, or merely reflects the humorlessness of a court where the teenaged Princess Elizabeth and Lady Jane Grey "forbid their servants to smile, lest the sin destroy their souls," is not clear. At any rate, the humor practiced in ordinary life also works to the displaced prince's advantage. Edward's manners may be subjected to blatant satire — "Ho, swine, slaves, pensioners of his grace's princely father, where be your manners?" responds the first boy he addresses with royal condescension; "Down on your marrow bones, all of ye, and do reverence to his kingly port and royal rags" — but the prince's survival, when he insists on the prerogatives of royalty among those who assume him to be a fellow pauper, turns out to hang implausibly on his protector's willingness to treat apparent insults as jokes and eccentricities. The exiled nobleman Miles Hendon, without compromising his morality, has long since adjusted his own manners to suit the reduced environment in which he now lives. In spite of this, or perhaps in nostalgic admiration of the haughtier life, he exhibits a saintly tolerance when the boy he generously befriends starts ordering him around.

Edward's response to a hospitable offer is "Prithee pour the water, and make not so many words!" Yet the reaction is a mild "Lo, the poor thing's madness is up with the time!" after which the derided host decides that he is "pleased with the jest" and willing to assume the indicated posture as the boy's inferior.

The reader, however, cannot help noticing that the prince, with his rigid insistence on protocol, has turned offensive. Even when one remembers that Edward can hardly be expected to take lightly the question of the sacredness of his person and position, it is difficult to escape the feeling that his insistence that the man who keeps saving his life show him constant obeisance is unpleasant. Etiquette itself, like a just system of law, requires that intent be taken into serious consideration in judging violations. Therefore, within the code of manners, mercilessly returning generosity of heroic proportions with petty reprimands is unspeakably rude.

In the end, the prince finds protocol used against him. As a presumed intruder who is thought impudent to royalty when he forces his way before the ersatz prince, Edward is made to realize that he is dependent for protection not on the enforcement of royalty-respecting law but on the virtue of Tom. A rule dear to the royal heart, "Up, thou mannerless clown! Wouldst sit in the presence of the king?"—which Edward has been trying to enforce on others throughout the book—does not sound quite so reasonable when directed at himself. The valuable point is thus made that rote etiquette, like law untempered with justice, is dangerous when not accompanied by the judgment to adjust it for motivation and circumstance.

Of course, one has to judge Edward mercifully, as well. One could say that a crown prince in a society that believes in the divine right of kings might have a handicap when it comes to learning to consider the feelings of others. At least until recently, royalty did not noticeably worry about the popularity polls, however many examples history offers that this might have been a good idea.

When Tom is catapulted into his new position, he has much less trouble adapting his manners to changed circumstances, and not only because a court etiquette book is available, whereas there was no such introduction to the etiquette of the streets. Unlike Edward, Tom seems to know from the beginning that etiquette is dependent on context. Therefore, he actually has less to learn: he has only to change his surface behavior, while the prince needs — but fails — to learn how manners operate as a force in society.

Tom's sophistication in this matter seems to arise from a healthy (and suspiciously American) sense of the possibility — however farfetched the reality may be — of social mobility. In order even to feel etiquette anxiety — which Tom does but Edward does not — one needs to know that behavior is different at different levels of society, and to believe that it is worth learning the practices of higher levels because one might be able to leave one's native circumstances and move up. (Moving downward also requires that understanding, if one is to be accepted and favorably judged by one's new peers, but the prospect is less eagerly studied or embraced.)

This assumption has less to do with the Tudor period, in which the book is set, than with the Victorian period, in which it was written. The anachronistic tip-off is Tom's recurrent anxiety about dining implements: "Poor Tom ate with his fingers mainly," and feared "the ordeal of dining all by himself with a multitude of curious eyes fastened upon him and a multitude of mouths whispering comments upon his performance — and upon his mistakes, if he should be so unlucky as to make any."

The fear of exposing social ineptitude by "using the wrong fork" is a quintessential Victorian preoccupation, oddly surviving into our own time, when the nineteenth century's sudden proliferation of specialized flatware has vanished. It could hardly have existed before, however. At this sixteenth-century royal court, the fork was all but unknown. Henry VIII "ate with his fingers mainly," too — a practice that has been faithfully depicted in twentieth-century films to demonstrate that he was a glutton with no control of his appetites, although it was proper procedure at the time. A century later, even the most finicky of monarchs, such as Louis XIV, in whose court the word "etiquette" in its present sense came into use, were still eating with their fingers mainly.

Another etiquette concept, periodically popular but especially exciting to those involved in the social upheavals of the Industrial Revolution, is even more significant to the story of prince and pauper. That is the idea that a gentleman may be defined by his behavior rather than his birth. When William of Wykeham, founder of Winchester College and New College at Oxford, articulated this idea in the fourteenth century as "Manners maketh man," it suggested that gentlemen — which is to say, men of gentle birth — should behave themselves, not that gentle behavior could earn one the birthright status of gentleman. It takes on new meaning, however, whenever an important change in the economic system raises the possibility of genuine social mobility.

Of course, money, not gentility, was, as it usually is, the major requirement for improving one's social fortunes in the nineteenth century. The connection between the two was a great deal more sordid than suggested by the rush to manners of that period. Etiquette, in one of its least pleasant aspects, was blatantly being brandished by those of high but decaying circumstances as a weapon against the social advancement of the new rich. True principles of manners were jettisoned as the inside knowledge of details of behavior — often deliberately complicated forms, freshly devised for use as social markers — was trickily employed as a test of eligibility.

However viciously used, this ploy — ever with us as groups at all levels seek to distinguish insiders from outsiders — is always ultimately unsuccessful against financial reality. Moneymakers, familiar with the combative use (or rather, misuse) of etiquette from their own efforts to distinguish themselves from their circles of origin, either learn the new rules or, at the least, provide their children with the leisure to do so. These dynamics remain familiar as a staple of the Victorian novel, where aristocratic but financially embarrassed parents are finally persuaded to admire and accept the industrial heiress their son coincidentally loves, because even they have to admit that she is beautifully behaved, in spite of the parents they deem so crude. At the end, everyone is happy — the young lovers, who have each other; the bride's parents, who gain rank for their descendants (whom they are expected to enjoy from a distance); and the bridegroom's parents, who get their hands on that despised money.

This sort of thing was going on in America, too, as well as between rich Americans and financially strapped European nobles. As befits an egalitarian society, however, America maintained the ideal that the moral principles underlying etiquette (along with a few virtues outside the domain of etiquette, notably hard work and American ingenuity) counted for everything, and inherited advantages for nothing.

The Prince and the Pauper is very much of this way of thinking. We are given to understand that the son of a thief, because he has a kind and honest heart, can make an easy transition not just to the very top of society but above it, convincingly, if temporarily, as its ruler. History to the contrary, this presumes that moral merit is the key qualification for the job.

Omitting our historical knowledge that the real Edward VI would not live through another year, the story ends on the promise that he is bound to be a king of stature. From the start, it is demonstrated that he has instincts above his origins, as it were. Not only does he rescue Tom from the palace guard and invite him for a visit, as if he were an equal, but he thoughtfully protects his visitor from scrutiny — again so that Tom will not be ashamed of his table manners: "The prince, with princely delicacy and breeding, sent away the servants so that his humble guest might not be embarrassed by their critical presence."

All this indicates a deep feeling for the principles of manners, especially the classic touchstone of hospitality, which also figures in the major religions as a moral test. When a god appears disguised as a pauper, those who turn him away will be punished and those who take him in rewarded. Excuses about who they thought he was or how little they had to share make no difference. Under the surface rules of the situation in this novel, the ejection of a scruffy intruder who has penetrated the security of the palace would be routine and reasonable, but Edward will not allow such a thing to happen. Later, on his own behalf, he keeps a running account of whom to reward and punish, based on how they treated him when his identity was not known.

Notwithstanding the prince's devotion to this principle of courtesy, the suspicion is bound to arise that it is the restrictions of court etiquette that have prevented him from understanding and feeling the life of his own realm, with all the tragedies and injustices to which his eyes are finally opened. His royalty itself stands in his way, and he becomes worthy of the throne only when he has left the protected environment that has shielded him from the complexities of life, and has been exposed to Tom Canty's world. The author is taking the position that a true king must be a man of the people — which, however politically laudable, is puzzling in the context of monarchy.

This populist proposition seems to be confirmed by the reaction of the sensible Tom when he is first exposed to court etiquette, and finds it not only bewildering, but stifling, ridiculous and boring. Remember that this is the boy who has admired and aped royalty all his life, now, on first contact, rejecting its ways because he realizes that they are not as straightforward as those of the simple life he found so distasteful. Immediately on finding himself living his dream of becoming a prince, he yearns to escape; and so he does at the end, when he happily turns the realm back to its true owner. The implication is that no right-thinking person would stand for such an elaborate life if not condemned to it by birth.

Something odd happens in between, however. Tom begins to get a taste for the royal way of life — and not just for the adulation directed toward his person, but for the ritual he at first misjudged because it was not practical; not just for the glamour, but for the power to do good. Neither he nor anyone reading about his pre-palace existence is likely to romanticize the slums as being more vital or authentic. The temptation to hang on at the top, perhaps with the rationalization that he would be doing so for the sake of what he could accomplish for the people, might have felled a less virtuous boy.

Having denied his mother, Tom has it within his power to deny the true claimant and make the throne his forever. He is ultimately too good to do this. Recoiling from that first dastardly act, he cannot do the second.

Was he in danger of being corrupted by the surrounding web of privilege from which the real prince's sabbatical saved him? Was Edward so ensnared in complicated etiquette that it would have stifled his innate goodness but for his adventure? Is etiquette actually the villain of the novel, threatening to blind the boys to pity and social justice? Have they both had lucky escapes — Edward temporarily, Tom permanently — from a force that would have killed their inborn sense of decency and humanity?

To think so requires assuming that a king is restricted by etiquette. *The Prince and the Pauper* amply illustrates not only that etiquette also pertains to all levels of society, perhaps more rigidly and certainly more dangerously at the bottom than at the top, but that kingship, if anything, frees one from etiquette, and in ways that are not good for the soul.

Far from being bound by the principles of manners, a secure prince is free to expect courtesies he has no intention of bestowing, and to operate without fear or consideration of the feelings of others. He is also free, at will or whim, to break or change any rules of etiquette, even his own, prevailing in his own court. Bound by etiquette? He is the one person who can eschew it with impunity, because he is above it.

That is what is corrupting. Edward, exposed to etiquette but never restrained by it, must have his behavior judged by others before he can understand the troubles of ordinary people who must live at the sufferance of one another and of their king. It is only when Tom experiences what it is to be above the laws of manners that he can test his own virtue against temptation. And so, as the pauper acquires a whiff of polish and the prince acquires a touch of humility, a common ideal standard emerges, in which good-heartedness, through study and experience, is honed into something socially useful.

THE PRINCE

AND THE PAUPER

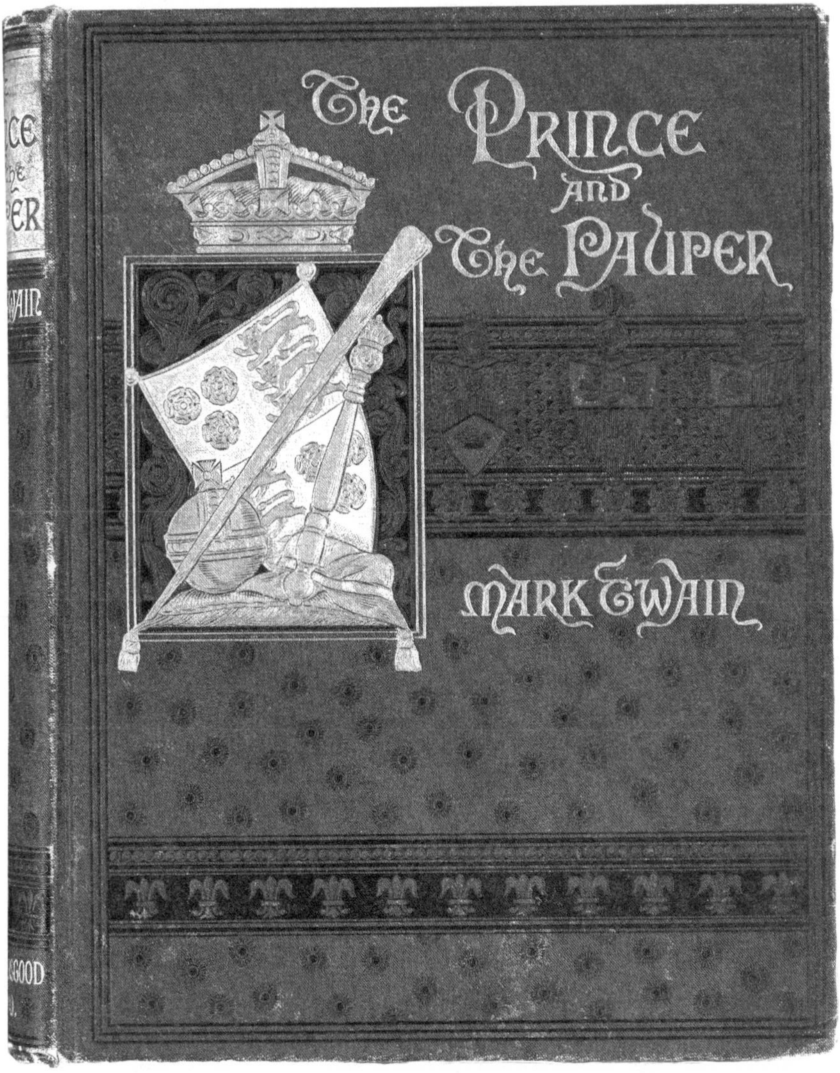
THE PRINCE AND THE PAUPER
MARK TWAIN

HUGH LATIMER, *Bishop of Worcester*, to LORD CROMWELL, *on the birth of the* PRINCE OF WALES (*afterward* EDWARD VI.).

FROM THE NATIONAL MANUSCRIPTS PRESERVED BY THE BRITISH GOVERNMENT.

Ryght honorable Salutem in Christo Jesu. And syr here ys no lesse joyynge and rejoysynge in thes partees for the byrth of our prynce, hoom we hungurde for so longe, then ther was (I trow) inter vicinos att the byrth of S. J. Baptyste. As thys berer master Evance can tell you. Gode gyffe us alle grace to yelde dew thankes to our Lorde gode, gode of Inglonde, ffor verely he hath shoyd hym selff gode of Inglonde or rather an Inglyssh gode, yf we consydyr and pondyr welle alle hys procedynges with us from tyme to tyme. He hath overcumme alle our yllnesse with hys excedynge goodnesse, so that we ar now moor then compellyd to serve hym, seke hys glory, promott hys wurde, yf the devylle of alle devyllys be natt in us. We have now the stooppe of vayne trustes, and the stey of vayne expectations, lett us alle pray for hys preservatione. And I for my partt wylle wyssh that hys grace allways have and evyn now from the begynnynge governares, instructores and offyceres of ryght jugmente, ne optimum ingenium non optima educatione depravetur. Butt whatt a grett fowlle am I! So whatt devotione shoyth many tymys butt lytelle dyscretione. And thus the gode of Inglonde be ever with you in alle your procedynges.

The xix of oct.

Youres H. L. B. of Wurcester

now att Hartlebury.

Yf you wolde excytt thys berar to be moor hartye ayen the abuse of ymagry or mor forwarde to promotte the veryte, ytt myght doo goode. Natt that ytt came of me, butt of your selffe etc.

Hugh Latimer, *Bishop of Worcester*, to Lord Cromwell, *on the birth of the* Prince of Wales (*afterward* Edward VI.).

FROM THE NATIONAL MANUSCRIPTS PRESERVED BY THE BRITISH GOVERNMENT.

Ryght honorable, *Salutem in Christo Jesu*, and Syr here ys no lesse joynge and rejossynge in thes partees for the byrth of our prynce, hoom we hungurde for so longe, then ther was (I trow), *inter vicinos* att the byrth of S. I. Baptyste, as thys berer, Master Erance, can telle you. Gode gyffe us alle grace, to yelde dew thankes to our Lorde Gode, Gode of Inglonde, for verely He hathe shoyd Hym selff Gode of Inglonde, or rather an Inglyssh Gode, yf we consydyr and pondyr welle alle Hys procedynges with us from tyme to tyme. He hath overcumme alle our yllnesse with Hys excedynge goodnesse, so that we ar now moor then compellyd to serve Hym, seke Hys glory, promott Hys wurde, yf the Devylle of alle Devylles be natt in us. We have now the stooppe of vayne trustes ande the stey of vayne expectations; lett us alle pray for hys preservatione. Ande I for my partt wylle wyssh that hys Grace allways have, and evyn now from the begynynge, Governares, Instructores and offyceres of ryght jugmente, *ne optimum ingenium non optimâ educatione depravetur*.

Butt whatt a grett fowlle am I! So, whatt devotione shoyth many tymys butt lytelle dyscretione! Ande thus the Gode of Inglonde be ever with you in alle your procedynges.

The 19 of October.

Youres, H. L. B. of Wurcestere, now att Hartlebury.

Yf you wolde excytt thys berere to be moore hartye ayen the abuse of ymagry or mor forwarde to promotte the veryte, ytt myght doo goode. Natt that ytt came of me, butt of your selffe, &c.

(*Addressed*) To the Ryght Honorable Loorde P. Sealle hys synguler gode Lorde.

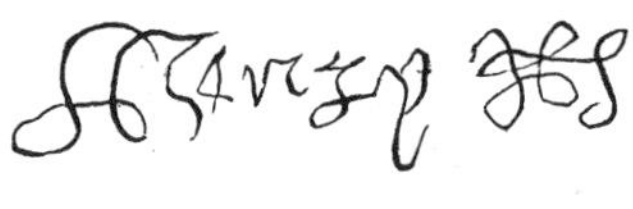

THE

PRINCE AND THE PAUPER

A TALE

FOR YOUNG PEOPLE OF ALL AGES

BY

MARK TWAIN

WITH ONE HUNDRED AND NINETY-TWO ILLUSTRATIONS

BOSTON

JAMES R. OSGOOD AND COMPANY

1882

FRANKLIN PRESS:

ELECTROTYPED AND PRINTED BY RAND, AVERY, AND COMPANY,

BOSTON.

TO

THOSE GOOD-MANNERED AND AGREEABLE CHILDREN,

SUSIE AND CLARA CLEMENS,

This Book

IS AFFECTIONATELY INSCRIBED

BY THEIR FATHER.

THE quality of mercy . . .
                    is twice bless'd;
It blesseth him that gives, and him that takes;
'Tis mightiest in the mightiest: it becomes
The thronéd monarch better than his crown.

Merchant of Venice.

CONTENTS.

LIST OF ILLUSTRATIONS.

I WILL set down a tale as it was told to me by one who had it of his father, which latter had it of *his* father, this last having in like manner had it of *his* father — and so on, back and still back, three hundred years and more, the fathers transmitting it to the sons and so preserving it. It may be history, it may be only a legend, a tradition. It may have happened, it may not have happened: but it *could* have happened. It may be that the wise and the learned believed it in the old days; it may be that only the unlearned and the simple loved it and credited it.

THE BIRTH OF
THE PRINCE AND THE PAUPER

THE PRINCE AND THE PAUPER.

CHAPTER I.

THE BIRTH OF THE PRINCE AND THE PAUPER.

"SPLENDID PAGEANTS AND GREAT BONFIRES."

In the ancient city of London, on a certain autumn day in the second quarter of the sixteenth century, a boy was born to a poor family of the name of Canty, who did not want him. On the same day another English child was born to a rich family of the name of Tudor, who did want him. All England wanted him too. England had so longed for him, and hoped for him, and prayed God for him, that, now that he was really come, the people went nearly mad for joy. Mere acquaintances hugged and kissed each other and cried. Everybody took a holiday;

and high and low, rich and poor, feasted and danced and sang, and got very mellow; and they kept this up for days and nights together. By day, London was a sight to see, with gay banners waving from every balcony and housetop, and splendid pageants marching along. By night, it was again a sight to see, with its great bonfires at every corner, and its troops of revellers making merry around them. There was no talk in all England but of the new baby, Edward Tudor, Prince of Wales, who lay lapped in silks and satins, unconscious of all this fuss, and not knowing that great lords and ladies were tending him and watching over him — and not caring, either. But there was no talk about the other baby, Tom Canty, lapped in his poor rags, except among the family of paupers whom he had just come to trouble with his presence.

TOM'S EARLY
LIFE

CHAPTER II.

TOM'S EARLY LIFE.

Let us skip a number of years.

London was fifteen hundred years old, and was a great town — for that day. It had a hundred thousand inhabitants — some think double as many. The streets were very narrow, and crooked, and dirty, especially in the part where Tom Canty lived, which was not far from London Bridge. The houses were of wood, with the second story projecting over the first, and the third sticking its elbows out beyond the second. The higher the houses grew, the broader they grew. They were skeletons of strong criss-cross beams, with solid material between, coated with plaster. The beams were painted red or blue or black, according to the owner's taste, and this gave the houses a very picturesque look. The windows were small, glazed with little diamond-shaped panes, and they opened outward, on hinges, like doors.

The house which Tom's father lived in was up a foul little pocket called Offal Court, out of Pudding Lane. It was small, decayed, and rickety, but it was packed full of wretchedly poor families. Canty's tribe occupied a room on the third floor. The mother and father had a sort of bedstead in the corner; but Tom, his grandmother, and his two sisters, Bet and Nan, were not restricted — they had all the floor to themselves, and might sleep where they chose. There were the remains of a blanket or two, and some bundles of ancient and dirty straw, but these could not rightly be called beds, for they were not organized; they were kicked into a general pile, mornings, and selections made from the mass at night, for service.

OFFAL COURT.

Bet and Nan were fifteen years old — twins. They were good-hearted girls, unclean, clothed in rags, and profoundly ignorant. Their mother was like them. But the father and the grandmother were a couple of fiends. They got drunk whenever they could; then they fought each other or anybody else who came in the way; they cursed and swore always, drunk or sober; John Canty was a thief, and his mother a beggar. They made beggars of the children, but failed to make thieves of them. Among, but not of, the dreadful rabble that inhabited the house, was a good old priest whom the King had turned out of house and home with a pension of a few farthings, and he used to get the children aside and teach them right ways secretly. Father Andrew also taught Tom a little Latin, and how to read and write; and would have done the same with the girls, but they were afraid of the jeers of their friends, who could not have endured such a queer accomplishment in them.

All Offal Court was just such another hive as Canty's house. Drunkenness, riot and brawling were the order, there, every night and nearly all night long. Broken heads were as common as hunger in that place. Yet little Tom was not unhappy. He had a hard time of it, but did not know it. It was the sort of time that all the Offal Court boys had, therefore he supposed it was the correct and comfortable thing. When he came home empty handed at night, he knew his father would curse him and thrash him first, and that when he was done the awful grandmother would do it all over again and improve on it; and that away in the night his starving mother would slip to him stealthily with any miserable scrap or crust she had been able to save for him by going hungry herself, notwithstanding she was often caught in that sort of treason and soundly beaten for it by her husband.

"WITH ANY MISERABLE CRUST."

No, Tom's life went along well enough, especially in summer. He only begged just enough to save himself, for the laws against mendicancy were stringent, and the penalties heavy; so he put in a good deal of his time listening to good Father Andrew's charming old tales and legends about giants and fairies, dwarfs and genii, and enchanted castles, and gorgeous kings and princes. His head grew to be full of these wonderful things, and many a night as he lay in the dark on his scant and offensive straw, tired, hungry, and smarting from a thrashing, he unleashed his imagination and soon forgot his aches and pains in delicious picturings to himself of the charmed life of a petted prince in a regal palace. One desire came in time to haunt him day and night: it was to see a real prince, with his own eyes. He spoke of it once to some of his Offal Court comrades; but they jeered him and scoffed him so unmercifully that he was glad to keep his dream to himself after that.

"HE OFTEN READ THE PRIEST'S BOOKS."

He often read the priest's old books and got him to explain and enlarge upon them. His dreamings and readings worked certain changes in him, by and by. His dream-people were so fine that he grew to lament his shabby clothing and his dirt, and to wish to be clean and better clad. He went on playing in the mud just the same, and enjoying it, too; but instead of

"SAW POOR ANNE ASKEW BURNED."

splashing around in the Thames solely for the fun of it, he began to find an added value in it because of the washings and cleansings it afforded. Tom could always find something going on around the Maypole in Cheapside, and at the fairs; and now and then he and the rest of London had a chance to see a military parade when some famous unfortunate was carried prisoner to the Tower, by land or boat. One summer's day he saw poor Anne Askew and three men burned at the stake in Smithfield, and heard an ex-Bishop preach a sermon to them which did not interest him. Yes, Tom's life was varied and pleasant enough, on the whole.

By and by Tom's reading and dreaming about princely life wrought such a strong effect upon him that he began to *act* the prince, unconsciously. His speech and manners became curiously ceremonious and courtly, to the vast admiration and amusement of his intimates. But Tom's influence among these young people began to grow, now, day by day; and in time he came to be looked up to, by them, with a sort of wondering awe, as a superior being. He seemed to know so much! and he could do and say such marvellous things! and withal, he was so deep and wise! Tom's remarks, and Tom's performances, were reported by the boys to their elders; and these, also, presently began to discuss Tom Canty, and to regard him as a most gifted and extraordinary creature. Full grown people brought their perplexities to Tom for solution, and were often astonished at the wit and wisdom of his decisions. In fact he was become a hero to all who knew him except his own family—these, only, saw nothing in him.

"BROUGHT THEIR PERPLEXITIES TO TOM."

Privately, after a while, Tom organized a royal court! He was the prince; his special comrades were guards, chamberlains, equerries, lords and ladies in waiting, and the royal family. Daily the mock prince was received with elaborate ceremonials borrowed by Tom from his romantic readings; daily the great affairs of the mimic kingdom were discussed in the royal council, and daily his mimic highness issued decrees to his imaginary armies, navies, and viceroyalties.

After which, he would go forth in his rags and beg a few farthings, eat his poor crust, take his customary cuffs and abuse, and then stretch himself upon his handful of foul straw, and resume his empty grandeurs in his dreams.

And still his desire to look just once upon a real prince, in the flesh, grew upon him, day by day, and week by week, until at last it absorbed all other desires, and became the one passion of his life.

"LONGING FOR THE PORK-PIES."

One January day, on his usual begging tour, he tramped despondently up and down the region round about Mincing Lane and Little East Cheap, hour after hour, bare-footed and cold, looking in at cook-shop windows and longing for the dreadful pork-pies and other deadly inventions displayed there — for to him these were dainties fit for the angels; that is, judging by the smell, they were — for it had never been his good luck to own and eat one. There was a cold drizzle of rain; the atmosphere was murky; it was a melancholy day. At night Tom reached home so wet and tired and hungry that it was not possible for his father and grandmother to observe his forlorn condition and not be moved — after their fashion; wherefore they gave him a brisk cuffing at once and sent him to bed. For a long time his pain and hunger, and the swearing and fighting going on in the building, kept him awake; but at last his thoughts drifted away to far, romantic lands, and he fell asleep in the company of jewelled and gilded princelings who lived in vast palaces, and had servants salaaming before them or flying to execute their orders. And then, as usual, he dreamed that *he* was a princeling himself.

All night long the glories of his royal estate shone upon him, he moved among great lords and ladies, in a blaze of light, breathing perfumes, drinking in delicious music, and answering the reverent obeisances of the glittering throng as it parted to make way for him, with here a smile, and there a nod of his princely head.

And when he awoke in the morning and looked upon the wretchedness about him, his dream had had its usual effect — it had intensified the sordidnesss of his surroundings a thousand fold. Then came bitterness, and heart-break, and tears.

TOM'S
MEETING WITH
THE
PRINCE

CHAPTER III.

TOM'S MEETING WITH THE PRINCE.

TOM got up hungry, and sauntered hungry away, but with his thoughts busy with the shadowy splendors of his night's dreams. He wandered here and there in the city, hardly noticing where he was going, or what was happening around him. People jostled him, and some gave him rough speech; but it was all lost on the musing boy. By and by he found himself at Temple Bar, the farthest from home he had ever travelled in that direction. He stopped and considered a moment, then fell into his imaginings again, and passed on outside the walls of London. The Strand had ceased to be a country-road then, and regarded itself as a street, but

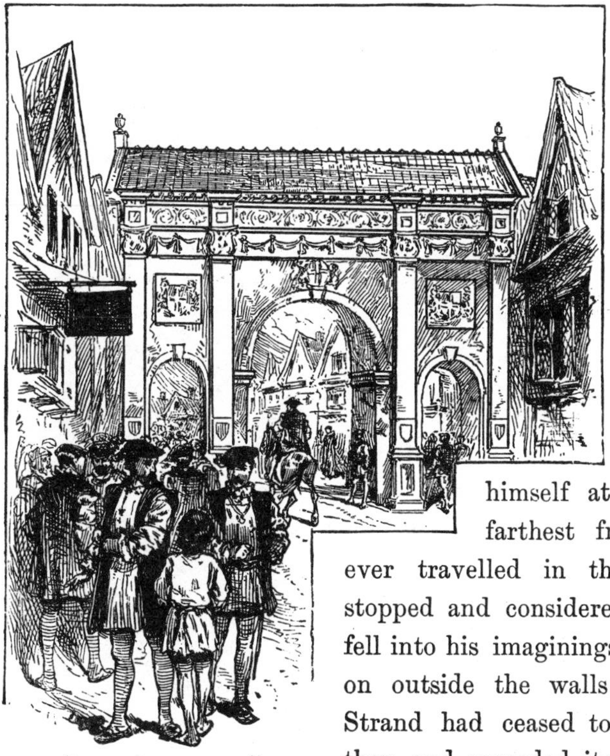

"AT TEMPLE BAR."

by a strained construction; for, though there was a tolerably compact row of houses on one side of it, there were only some scattering great buildings on the other, these being palaces of rich nobles, with ample and beautiful grounds stretching to the river, — grounds that are now closely packed with grim acres of brick and stone.

Tom discovered Charing Village presently, and rested himself at the beautiful cross built there by a bereaved king of earlier days; then idled down a quiet, lovely road, past the great cardinal's stately palace, toward a far more mighty and majestic palace beyond, — Westminster. Tom stared in glad wonder at the vast pile of masonry, the wide-spreading wings, the frowning bastions and turrets, the huge stone gateway, with its gilded bars and its magnificent array of colossal granite lions, and other the signs and symbols of English royalty. Was the desire of his soul to be satisfied at last? Here, indeed, was a king's palace. Might he not hope to see a prince now, — a prince of flesh and blood, if Heaven were willing?

At each side of the gilded gate stood a living statue, that is to say, an erect and stately and motionless man-at-arms, clad from head to heel in shining steel armor. At a respectful distance were many country folk, and people from the city, waiting for any chance glimpse of royalty that might offer. Splendid carriages, with splendid people in them and splendid servants outside, were arriving and departing by several other noble gateways that pierced the royal enclosure.

Poor little Tom, in his rags, approached, and was moving slow and timidly past the sentinels, with a beating heart and a rising hope, when all at once he caught sight through the golden bars of a spectacle that almost made him shout for joy. Within was a comely boy, tanned and brown with sturdy out-door sports and exercises, whose clothing was all of lovely silks and satins, shining with jewels; at his hip a little jewelled sword and dagger; dainty buskins on his feet, with red heels; and on his head a jaunty crimson cap, with drooping plumes fastened with a great sparkling gem. Several gorgeous gentlemen stood near, — his servants, without a doubt. Oh! he was a prince — a prince, a living prince, a real prince — without the shadow of a question; and the prayer of the pauper-boy's heart was answered at last.

Tom's breath came quick and short with excitement, and his eyes grew big with wonder and delight. Every thing gave way in his mind instantly to one desire: that was to get close to the prince, and have a good, devouring look at him. Before he knew what he was about, he had his face against the gate-bars. The next instant one of the soldiers snatched him rudely away, and sent him spinning among the gaping crowd of country gawks and London idlers. The soldier said, —

"Mind thy manners, thou young beggar!"

The crowd jeered and laughed; but the young prince sprang to the gate with his face flushed, and his eyes flashing with indignation, and cried out, —

"LET HIM IN!"

"How dar'st thou use a poor lad like that! How dar'st thou use the King my father's meanest subject so! Open the gates, and let him in!"

You should have seen that fickle crowd snatch off their hats then. You should have heard them cheer, and shout, "Long live the Prince of Wales!"

The soldiers presented arms with their halberds, opened the gates, and presented again as the little Prince of Poverty passed in, in his fluttering rags, to join hands with the Prince of Limitless Plenty.

Edward Tudor said, —

"Thou lookest tired and hungry: thou'st been treated ill. Come with me."

Half a dozen attendants sprang forward to — I don't know what; interfere, no doubt. But they were waved aside with a right royal gesture, and they stopped stock still where they were, like so many statues. Edward took Tom to a rich apartment in the palace, which he called his cabinet. By his command a repast was brought such as Tom had never encountered before except in books. The prince, with princely delicacy and breeding, sent away the servants, so that his humble guest might not be embarrassed by their critical presence; then he sat near by, and asked questions while Tom ate.

"What is thy name, lad?"

"Tom Canty, an' it please thee, sir."

"'Tis an odd one. Where dost live?"

"In the city, please thee, sir. Offal Court, out of Pudding Lane."

"Offal Court! Truly 'tis another odd one. Hast parents?"

"Parents have I, sir, and a grand-dam likewise that is but indifferently precious to me, God forgive me if it be offence to say it — also twin sisters, Nan and Bet."

"Then is thy grand-dam not over kind to thee, I take it."

"Neither to any other is she, so please your Worship. She hath a wicked heart, and worketh evil all her days."

"Doth she mistreat thee?"

"There be times that she stayeth her hand, being asleep or overcome with drink; but when she hath her judgment clear again, she maketh it up to me with goodly beatings."

A fierce look came into the little prince's eyes, and he cried out,—

"What! Beatings?"

"Oh, indeed, yes, please you, sir."

"HOW OLD BE THESE?"

"*Beatings!*—and thou so frail and little. Hark ye: before the night come, she shall hie her to the Tower. The King my father"—

"In sooth, you forget, sir, her low degree. The Tower is for the great alone."

"True, indeed. I had not thought of that. I will consider of her punishment. Is thy father kind to thee?"

"Not more than Gammer Canty, sir."

"Fathers be alike, mayhap. Mine hath not a doll's temper. He smiteth with a heavy hand, yet spareth me: he spareth me not always with his tongue, though, sooth to say. How doth thy mother use thee?"

"She is good, sir, and giveth me neither sorrow nor pain of any sort. And Nan and Bet are like to her in this."

"How old be these?"

"Fifteen, an' it please you, sir."

"The Lady Elizabeth, my sister, is fourteen, and the Lady Jane Grey, my cousin, is of mine own age, and comely and gracious withal; but my sister the Lady Mary, with her gloomy mien and— Look you: do thy sisters forbid their servants to smile, lest the sin destroy their souls?"

"They? Oh, dost think, sir, that *they* have servants?"

The little prince contemplated the little pauper gravely a moment, then said,—

"And prithee, why not? Who helpeth them undress at night? who attireth them when they rise?"

"None, sir. Wouldst have them take off their garment, and sleep without,—like the beasts?"

"Their garment! Have they but one?"

"Ah, good your worship, what would they do with more? Truly they have not two bodies each."

"It is a quaint and marvellous thought! Thy pardon, I had not meant to laugh. But thy good Nan and thy Bet shall have raiment and lackeys enow, and that soon, too: my cofferer shall look to it. No, thank me not; 'tis nothing. Thou speakest well; thou hast an easy grace in it. Art learned?"

"I know not if I am or not, sir. The good priest that is called Father Andrew taught me, of his kindness, from his books."

"Know'st thou the Latin?"

"But scantly, sir, I doubt."

"Learn it, lad: 'tis hard only at first. The Greek is harder; but neither these nor any tongues else, I think, are hard to the Lady Elizabeth and my cousin. Thou shouldst hear those damsels at it! But tell me of thy Offal Court. Hast thou a pleasant life there?"

"In truth, yes, so please you, sir, save when one is hungry. There be Punch-and-Judy shows, and monkeys, — oh, such antic creatures! and so bravely dressed! — and there be plays wherein they that play do shout and fight till all are slain, and 'tis so fine to see, and costeth but a farthing — albeit 'tis main hard to get the farthing, please your worship."

"DOFF THY RAGS, AND DON THESE SPLENDORS."

"Tell me more."

"We lads of Offal Court do strive against each other with the cudgel, like to the fashion of the 'prentices, sometimes."

The prince's eyes flashed. Said he, —

"Marry, that would not I mislike. Tell me more."

"We strive in races, sir, to see who of us shall be fleetest."

"That would I like also. Speak on."

"In summer, sir, we wade and swim in the canals and in the river, and each doth duck his neighbor, and spatter him with water, and dive and shout and tumble and" —

"'Twould be worth my father's kingdom but to enjoy it once! Prithee go on."

"We dance and sing about the Maypole in Cheapside; we play in the sand, each covering his neighbor up; and times we make mud pastry — oh the lovely mud, it hath not its like for delightfulness in all the world! — we do fairly wallow in the mud, sir, saving your worship's presence."

"Oh, prithee, say no more, 'tis glorious! If that I could but clothe me in raiment like to thine, and strip my feet, and revel in the mud once, just once, with none to rebuke me or forbid, meseemeth I could forego the crown!"

"And if that I could clothe me once, sweet sir, as thou art clad — just once" —

"Oho, wouldst like it? Then so shall it be. Doff thy rags, and don these splendors, lad! It is a brief happiness, but will be not less keen for that. We will have it while we may, and change again before any come to molest."

A few minutes later the little Prince of Wales was garlanded with Tom's fluttering odds and ends, and the little Prince of Pauperdom was tricked out in the gaudy plumage of royalty. The two went and stood side by side before a great mirror, and lo, a miracle: there did not seem to have been any change made! They stared at each other, then at the glass, then at each other again. At last the puzzled princeling said, —

"What dost thou make of this?"

"Ah, good your worship, require me not to answer. It is not meet that one of my degree should utter the thing."

"Then will *I* utter it. Thou hast the same hair, the same eyes, the same voice and manner, the same form and stature, the same face and countenance, that I bear. Fared we forth naked, there is none could say which was you, and which the Prince of Wales. And, now that I am clothed as thou wert clothed, it seemeth I should be able the more nearly to feel as thou didst when the brute soldier — Hark ye, is not this a bruise upon your hand?"

"Yes; but it is a slight thing, and your worship knoweth that the poor man-at-arms" —

"Peace! It was a shameful thing and a cruel!" cried the little prince, stamping his bare foot. "If the King — Stir not a step till I come again! It is a command!"

In a moment he had snatched up and put away an article of national importance that lay upon a table, and was out at the door and flying through the palace grounds in his bannered rags, with a hot face and glowing eyes. As soon as he reached the great gate, he seized the bars, and tried to shake them, shouting, —

"Open! Unbar the gates!"

The soldier that had maltreated Tom obeyed promptly; and as the prince burst through the portal, half-smothered with royal wrath, the soldier fetched him a sounding box on the ear that sent him whirling to the roadway, and said, —

"Take that, thou beggar's spawn, for what thou got'st me from his Highness!"

The crowd roared with laughter. The prince picked himself out of the mud, and made fiercely at the sentry, shouting, —

"I am the Prince of Wales, my person is sacred; and thou shalt hang for laying thy hand upon me!"

The soldier brought his halberd to a present-arms and said mockingly, —

"I salute your gracious Highness." Then angrily, "Be off, thou crazy rubbish!"

Here the jeering crowd closed around the poor little prince, and hustled him far down the road, hooting him, and shouting, "Way for his royal Highness! way for the Prince of Wales!"

"I SALUTE YOUR GRACIOUS HIGHNESS!"

THE PRINCE'S
TROUBLES
BEGIN

CHAPTER IV.

THE PRINCE'S TROUBLES BEGIN.

AFTER hours of persistent pursuit and persecution, the little prince was at last deserted by the rabble and left to himself. As long as he had been able to rage against the mob, and threaten it royally, and royally utter commands that were good stuff to laugh at, he was very entertaining; but when weariness finally forced him to be silent, he was no longer of use to his tormentors, and they sought amusement elsewhere. He looked about him, now, but could not recognize the locality. He was within the city of London — that was all he knew. He moved on, aimlessly, and in a little while the houses thinned, and the passers-by were infrequent. He bathed his bleeding feet in the brook which flowed then where Farringdon street now is; rested a few moments, then passed on, and presently came upon a great space with only a few scattered houses in it, and a prodigious church. He recognized this church. Scaffoldings were about, everywhere, and swarms of workmen; for it was undergoing elaborate repairs. The prince took heart at once — he felt that his troubles were at an end, now. He said to himself, "It is the ancient Grey Friars' church, which the king my father hath taken from the monks and given for a home forever for poor and forsaken children, and new-named it Christ's Church. Right gladly will they serve the son of him who hath done so generously by them — and the more that that son is himself as poor and as forlorn as any that be sheltered here this day, or ever shall be."

He was soon in the midst of a crowd of boys who were running, jumping, playing at ball and leap-frog and otherwise disporting themselves, and right noisily, too. They were all dressed alike, and in the fashion which in that day prevailed among serving-men and 'prentices[1] — that is to say, each had on the crown of his head a flat black cap about the size of a saucer, which was not useful as a covering, it being of such scanty dimensions, neither was it ornamental; from beneath it

"SET UPON BY DOGS."

the hair fell, unparted, to the middle of the forehead, and was cropped straight around; a clerical band at the neck; a blue gown that fitted closely and hung as low as the knees or lower; full sleeves; a broad red belt; bright yellow stockings, gartered above the knees; low shoes with large metal buckles. It was a sufficiently ugly costume.

[1] See Note 1, at end of the volume.

The boys stopped their play and flocked about the prince, who said with native dignity —

"Good lads, say to your master that Edward Prince of Wales desireth speech with him."

A great shout went up, at this, and one rude fellow said —

"Marry, art thou his grace's messenger, beggar?"

The prince's face flushed with anger, and his ready hand flew to his hip, but there was nothing there. There was a storm of laughter, and one boy said —

"Didst mark that? He fancied he had a sword — belike he is the prince himself."

This sally brought more laughter. Poor Edward drew himself up proudly and said —

"I am the prince; and it ill beseemeth you that feed upon the king my father's bounty to use me so."

This was vastly enjoyed, as the laughter testified. The youth who had first spoken, shouted to his comrades —

"Ho, swine, slaves, pensioners of his grace's princely father, where be your manners? Down on your marrow bones, all of ye, and do reverence to his kingly port and royal rags!"

With boisterous mirth they dropped upon their knees in a body and did mock homage to their prey. The prince spurned the nearest boy with his foot, and said fiercely —

"Take thou that, till the morrow come and I build thee a gibbet!"

Ah, but this was not a joke — this was going beyond fun. The laughter ceased on the instant, and fury took its place. A dozen shouted —

"Hale him forth! To the horse-pond, to the horse-pond! Where be the dogs? Ho, there, Lion! ho, Fangs!"

Then followed such a thing as England had never seen before —

the sacred person of the heir to the throne rudely buffeted by plebeian hands, and set upon and torn by dogs.

As night drew to a close that day, the prince found himself far down in the close-built portion of the city. His body was bruised, his hands were bleeding, and his rags were all besmirched with mud. He wandered on and on, and grew more and more bewildered, and so tired and faint he could hardly drag one foot after the other. He had ceased to ask questions of any one, since they brought him only insult instead of information. He kept muttering to himself, "Offal court—that is the name; if I can but find it before my strength is wholly spent and I drop, then am I saved—for his people will take me to the palace and prove that I am none of theirs, but the true prince, and I shall have mine own again." And now and then his mind reverted to his treatment by those rude Christ's Hospital boys, and he said, "When I am king, they shall not have bread and shelter only, but also teachings out of books; for a full belly is little worth where the mind is starved, and the heart. I will keep this diligently in my re-

"A DRUNKEN RUFFIAN COLLARED HIM."

membrance, that this day's lesson be not lost upon me, and my people suffer thereby; for learning softeneth the heart and breedeth gentleness and charity."[1]

The lights began to twinkle, it came on to rain, the wind rose, and a raw and gusty night set in. The houseless prince, the homeless heir to the throne of England, still moved on, drifting deeper into the maze of squalid alleys where the swarming hives of poverty and misery were massed together.

Suddenly a great drunken ruffian collared him and said —

"Out to this time of night again, and hast not brought a farthing home, I warrant me! If it be so, an' I do not break all the bones in thy lean body, then am I not John Canty, but some other."

The prince twisted himself loose, unconsciously brushed his profaned shoulder, and eagerly said —

"O, art *his* father, truly? Sweet heaven grant it be so — then wilt thou fetch him away and restore me!"

"*His* father? I know not what thou mean'st; I but know I am *thy* father, as thou shalt soon have cause to" —

"O, jest not, palter not, delay not! — I am worn, I am wounded, I can bear no more. Take me to the king my father, and he will make thee rich beyond thy wildest dreams. Believe me, man, believe me! — I speak no lie, but only the truth! — put forth thy hand and save me! I am indeed the Prince of Wales!"

The man stared down, stupefied, upon the lad, then shook his head and muttered —

"Gone stark mad as any Tom o' Bedlam!" — then collared him once more, and said with a coarse laugh and an oath, "But mad or no mad, I and thy Gammer Canty will soon find where the soft places in thy bones lie, or I'm no true man!"

With this he dragged the frantic and struggling prince away, and disappeared up a front court followed by a delighted and noisy swarm of human vermin.

[1] See Note 2, at end of the volume.

TOM AS A
PATRICIAN

CHAPTER V.

TOM AS A PATRICIAN.

TOM CANTY, left alone in the prince's cabinet, made good use of his opportunity. He turned himself this way and that before the great mirror, admiring his finery; then walked away, imitating the prince's high-bred carriage, and still observing results in the glass. Next he drew the beautiful sword, and bowed, kissing the blade, and laying it across his breast, as he had seen a noble knight do, by way of salute to the lieutenant of the Tower, five or six weeks before, when delivering the great lords of Norfolk and Surrey into his hands for captivity. Tom played with the jewelled dagger that hung upon his thigh; he examined the costly and exquisite ornaments of the room; he tried each of the sumptuous chairs, and thought how proud he would be if

"NEXT HE DREW THE SWORD."

the Offal Court herd could only peep in and see him in his grandeur. He wondered if they would believe the marvellous tale he should tell when he got home, or if they would shake their heads, and say his overtaxed imagination had at last upset his reason.

At the end of half an hour it suddenly occurred to him that the prince was gone a long time; then right away he began to feel lonely;

"RESOLVED TO FLY."

very soon he fell to listening and longing, and ceased to toy with the pretty things about him; he grew uneasy, then restless, then distressed. Suppose some one should come, and catch him in the prince's clothes, and the prince not there to explain. Might they not hang him at once, and inquire into his case afterward? He had heard that the great were prompt about small matters. His fears rose higher and higher; and trembling he softly opened the door to the antechamber, resolved to fly and seek the prince, and, through him, protection and release. Six gorgeous gentlemen-servants and two young pages of high degree, clothed like butterflies, sprung to their feet, and bowed low before him. He stepped quickly back, and shut the door. He said, —

"Oh, they mock at me! They will go and tell. Oh! why came I here to cast away my life?"

He walked up and down the floor, filled with nameless fears, listening, starting at every trifling sound. Presently the door swung open, and a silken page said, —

"The Lady Jane Grey."

"THE BOY WAS ON HIS KNEES."

The door closed, and a sweet young girl, richly clad, bounded toward him. But she stopped suddenly, and said in a distressed voice, —

"Oh, what aileth thee, my lord?"

Tom's breath was nearly failing him; but he made shift to stammer out, —

"Ah, be merciful, thou! In sooth I am no lord, but only poor Tom Canty of Offal Court in the city. Prithee let me see the prince, and he will of his grace restore to me my rags, and let me hence unhurt. Oh, be thou merciful, and save me!"

By this time the boy was on his knees, and supplicating with his eyes and uplifted hands as well as with his tongue. The young girl seemed horror-stricken. She cried out —

"O my lord, on thy knees? — and to *me!*"

Then she fled away in fright; and Tom, smitten with despair, sank down, murmuring —

"There is no help, there is no hope. Now will they come and take me."

Whilst he lay there benumbed with terror, dreadful tidings were speeding through the palace. The whisper, for it was whispered always, flew from menial to menial, from lord to lady, down all the long corridors, from story to story, from saloon to saloon, "The prince hath gone mad, the prince hath gone mad!" Soon every saloon, every marble hall, had its groups of glittering lords and ladies, and other groups of dazzling lesser folk, talking earnestly together in whispers, and every face had in it dismay. Presently a splendid official came marching by these groups, making solemn proclamation. —

"In the Name of the King!

Let none list to this false and foolish matter, upon pain of death, nor discuss the same, nor carry it abroad. In the name of the King!"

The whisperings ceased as suddenly as if the whisperers had been stricken dumb.

Soon there was a general buzz along the corridors, of "The prince See, the prince comes!"

Poor Tom came slowly walking past the low-bowing groups, trying to bow in return, and meekly gazing upon his strange surroundings with bewildered and pathetic eyes. Great nobles walked upon each side of him, making him lean upon them, and so steady his steps. Behind him followed the court-physicians and some servants.

Presently Tom found himself in a noble apartment of the palace, and heard the door close behind him. Around him stood those who had come with him.

"GREAT NOBLES WALKED UPON EACH SIDE OF HIM."

Before him, at a little distance, reclined a very large and very fat man, with a wide, pulpy face, and a stern expression. His large head was very gray; and his whiskers, which he wore only around his face, like a frame, were gray also. His clothing was of rich stuff, but old, and slightly frayed in places. One of his swollen legs had a pillow under it, and was wrapped in bandages. There was silence now; and there was no head there but was bent in reverence, except this man's. This stern-countenanced invalid was the dread Henry VIII. He said, — and his face grew gentle as he began to speak, —

"How now, my lord Edward, my prince? Hast been minded to cozen me, the good King thy father, who loveth thee, and kindly useth thee, with a sorry jest?"

"HE DROPPED UPON HIS KNEES."

Poor Tom was listening, as well as his dazed faculties would let him, to the beginning of this speech; but when the words "me the good King" fell upon his ear, his face blanched, and he dropped as instantly upon his knees as if a shot had brought him there. Lifting up his hands, he exclaimed, —

"Thou the *King?* Then am I undone indeed!"

This speech seemed to stun the King. His eyes wandered from face to face aimlessly, then rested, bewildered, upon the boy before him. Then he said in a tone of deep disappointment, —

"Alack, I had believed the rumor disproportioned to the truth; but I fear me 'tis not so." He breathed a heavy sigh, and said in a gentle voice, "Come to thy father, child: thou art not well."

Tom was assisted to his feet, and approached the Majesty of England, humble and trembling. The King took the frightened face between his hands, and gazed earnestly and lovingly into it a while, as if seeking some grateful sign of returning reason there, then pressed the curly head against his breast, and patted it tenderly. Presently he said, —

"Dost not know thy father, child? Break not mine old heart; say thou know'st me. Thou *dost* know me, dost thou not?"

"Yea: thou art my dread lord the King, whom God preserve!"

"True, true — that is well — be comforted, tremble not so; there is none here would hurt thee; there is none here but loves thee. Thou art better now; thy ill dream passeth — is't not so? And thou knowest thyself now also — is't not so? Thou wilt not miscall thyself again, as they say thou didst a little while agone?"

"I pray thee of thy grace believe me, I did but speak the truth, most dread lord; for I am the meanest among thy subjects, being a pauper born, and 'tis by a sore mischance and accident I am here, albeit I was therein nothing blameful. I am but young to die, and thou canst save me with one little word. Oh speak it, sir!"

"Die? Talk not so, sweet prince — peace, peace, to thy troubled heart — thou shalt not die!"

Tom dropped upon his knees with a glad cry, —

"God requite thy mercy, oh my King, and save thee long to bless thy land!" Then springing up, he turned a joyful face toward the two lords in waiting, and exclaimed, "Thou heard'st it! I am not to die: the King hath said it!" There was no movement, save that all bowed with grave respect; but no one spoke. He hesitated, a little confused, then turned timidly toward the King, saying, "I may go now?"

"HE TURNED WITH JOYFUL FACE."

"Go? Surely, if thou desirest. But why not tarry yet a little? Whither wouldst go?"

Tom dropped his eyes, and answered humbly, —

"Peradventure I mistook; but I did think me free, and so was I moved to seek again the kennel where I was born and bred to misery, yet which harboreth my mother and my sisters, and so is home to me; whereas these pomps and splendors whereunto I am not used — oh, please you, sir, to let me go!"

The King was silent and thoughtful a while, and his face betrayed a growing distress and uneasiness. Presently he said, with something of hope in his voice,—

"Perchance he is but mad upon this one strain, and hath his wits unmarred as toucheth other matter. God send it may be so! We will make trial."

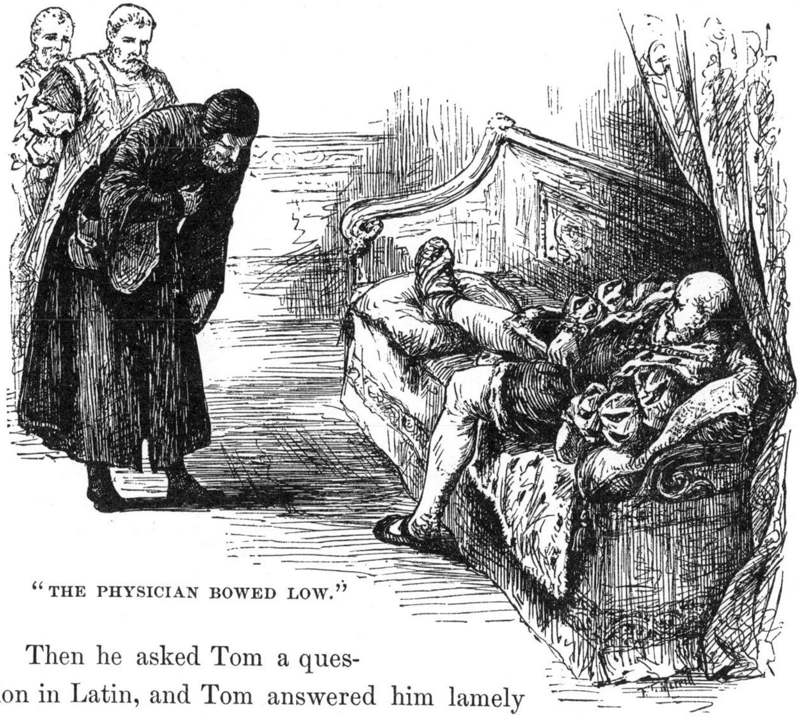

"THE PHYSICIAN BOWED LOW."

Then he asked Tom a question in Latin, and Tom answered him lamely in the same tongue. The King was delighted, and showed it. The lords and doctors manifested their gratification also. The King said,—

"'Twas not according to his schooling and ability, but sheweth that his mind is but diseased, not stricken fatally. How say you, sir?"

The physician addressed bowed low, and replied,—

"It jumpeth with mine own conviction, sire, that thou hast divined aright."

The King looked pleased with this encouragement, coming as it did from so excellent authority, and continued with good heart,—

"Now mark ye all: we will try him further."

He put a question to Tom in French. Tom stood silent a moment, embarrassed by having so many eyes centred upon him, then said diffidently,—

"I have no knowledge of this tongue, so please your majesty."

The King fell back upon his couch. The attendants flew to his assistance; but he put them aside, and said,—

"Trouble me not—it is nothing but a scurvy faintness. Raise me! there, 'tis sufficient. Come hither, child; there, rest thy poor troubled head upon thy father's heart, and be at peace. Thou'lt soon be well: 'tis but a passing fantasy. Fear thou not; thou'lt soon be well." Then he turned toward the company: his gentle manner changed, and baleful lightnings began to play from his eyes. He said,—

"List ye all! This my son is mad; but it is not permanent. Over-study hath done this, and somewhat too much of confinement. Away with his books and teachers! see ye to it. Pleasure him with sports, beguile him in wholesome ways, so that his health come again." He raised himself higher still, and went on with energy, "He is mad; but he is my son, and England's heir; and, mad or sane, still shall he reign! And hear ye further, and proclaim it: whoso speaketh of this his distemper worketh against the peace and order of these realms, and shall to the gallows! . . . Give me to drink—I burn: this sorrow sappeth my strength. . . . There, take away the cup. . . . Support me. There, that is well. Mad, is he? Were he a thousand times mad, yet is he Prince of Wales, and I the King will confirm it. This very morrow shall he be installed in his princely dignity in due and ancient form. Take instant order for it, my lord Hertford."

One of the nobles knelt at the royal couch, and said, —

"The King's majesty knoweth that the Hereditary Great Marshal of England lieth attainted in the Tower. It were not meet that one attainted" —

"THE KING FELL BACK UPON HIS COUCH."

"Peace! Insult not mine ears with his hated name. Is this man to live forever? Am I to be balked of my will? Is the prince to tarry uninstalled, because, forsooth, the realm lacketh an earl marshal free of treasonable taint to invest him with his honors? No, by the splendor of God! Warn my parliament to bring me Norfolk's doom before the sun rise again, else shall they answer for it grievously!"[1]

Lord Hertford said, —

"The King's will is law;" and, rising, returned to his former place.

[1] See Note 3, at end of the volume.

Gradually the wrath faded out of the old King's face, and he said, —

"Kiss me, my prince. There . . . what fearest thou? Am I not thy loving father?"

"Thou art good to me that am unworthy, O mighty and gracious lord: that in truth

"IS THIS MAN TO LIVE FOREVER?"

I know. But — but — it grieveth me to think of him that is to die, and" —

"Ah, 'tis like thee, 'tis like thee! I know thy heart is still the same, even though thy mind hath suffered hurt, for thou wert ever of a gentle spirit. But this duke standeth between thee and thine honors: I will have another in his stead that shall bring no taint to his great office. Comfort thee, my prince: trouble not thy poor head with this matter."

"But is it not I that speed him hence, my liege? How long might he not live, but for me?"

"Take no thought of him, my prince: he is not worthy. Kiss me once again, and go to thy trifles and amusements; for my malady distresseth me. I am aweary, and would rest. Go with thine uncle Hertford and thy people, and come again when my body is refreshed."

Tom, heavy-hearted, was conducted from the presence, for this last sentence was a death-blow to the hope he had cherished that now he would be set free. Once more he heard the buzz of low voices exclaiming, "The prince, the prince comes!"

His spirits sank lower and lower as he moved between the glittering files of bowing courtiers; for he recognized that he was indeed a captive now, and might remain forever shut up in this gilded cage, a forlorn and friendless prince, except God in his mercy take pity on him and set him free.

And, turn where he would, he seemed to see floating in the air the severed head and the remembered face of the great Duke of Norfolk, the eyes fixed on him reproachfully.

His old dreams had been so pleasant; but this reality was so dreary!

Tom
RECEIVES INSTRUCTIONS

CHAPTER VI.

TOM RECEIVES INSTRUCTIONS.

Tom was conducted to the principal apartment of a noble suite, and made to sit down — a thing which he was loath to do, since there were elderly men and men of high degree about him. He begged them to be seated, also, but they only bowed their thanks or murmured them, and remained standing. He would have insisted, but his "uncle" the earl of Hertford whispered in his ear —

"Prithee, insist not, my lord; it is not meet that they sit in thy presence."

The lord St. John was announced, and after making obeisance to Tom, he said —

"I come upon the king's errand, concerning a matter which requireth privacy. Will it please your royal highness to dismiss all that attend you here, save my lord the earl of Hertford?"

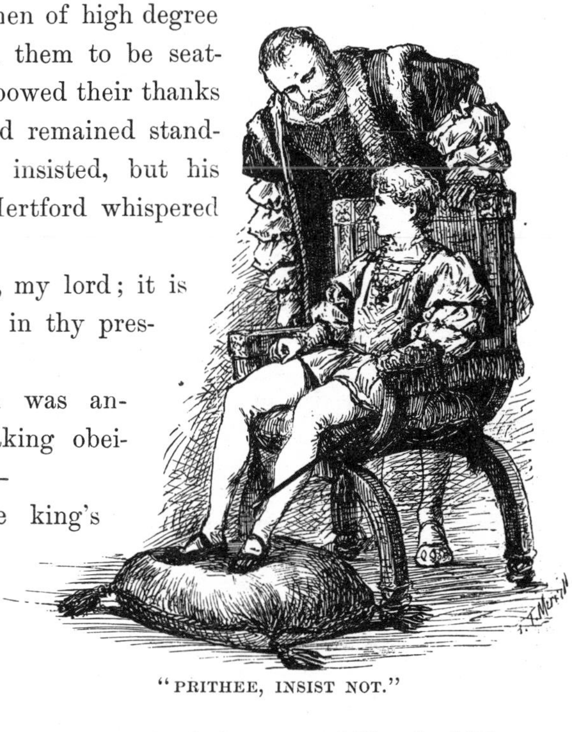

"PRITHEE, INSIST NOT."

Observing that Tom did not seem to know how to proceed, Hertford whispered him to make a sign with his hand and not trouble himself to speak unless he chose. When the waiting gentlemen had retired, lord St. John said —

"His majesty commandeth, that for due and weighty reasons of state, the prince's grace shall hide his infirmity in all ways that be within his power, till it be passed and he be as he was before. To wit, that he shall deny to none that he is the true prince, and heir to England's greatness; that he shall uphold his princely dignity, and shall receive, without word or sign of protest, that reverence and observance which unto it do appertain of right and ancient usage; that he shall cease to speak to any of that lowly birth and life his malady hath conjured out of the unwholesome imaginings of o'er-wrought fancy; that he shall strive with diligence to bring unto his memory again those faces which he was wont to know — and where he faileth he shall hold his peace, neither betraying by semblance of surprise, or other sign, that he hath forgot; that upon occasions of state, whensoever any matter shall perplex him as to the thing he should do or the utterance he should make, he shall show nought of unrest to the curious that look on, but take advice in that matter of the lord Hertford, or my humble self, which are commanded of the king to be upon this service and close at call, till this commandment be dissolved. Thus saith the king's majesty, who sendeth greeting to your royal highness and prayeth that God will of His mercy quickly heal you and have you now and ever in His holy keeping."

The lord St. John made reverence and stood aside. Tom replied, resignedly —

"The king hath said it. None may palter with the king's command, or fit it to his ease, where it doth chafe, with deft evasions. The king shall be obeyed."

Lord Hertford said —

"Touching the king's majesty's ordainment concerning books and such like serious matters, it may peradventure please your highness to ease your time with lightsome entertainment, lest you go wearied to the banquet and suffer harm thereby."

Tom's face showed inquiring surprise; and a blush followed when he saw lord St. John's eyes bent sorrowfully upon him. His lordship said—

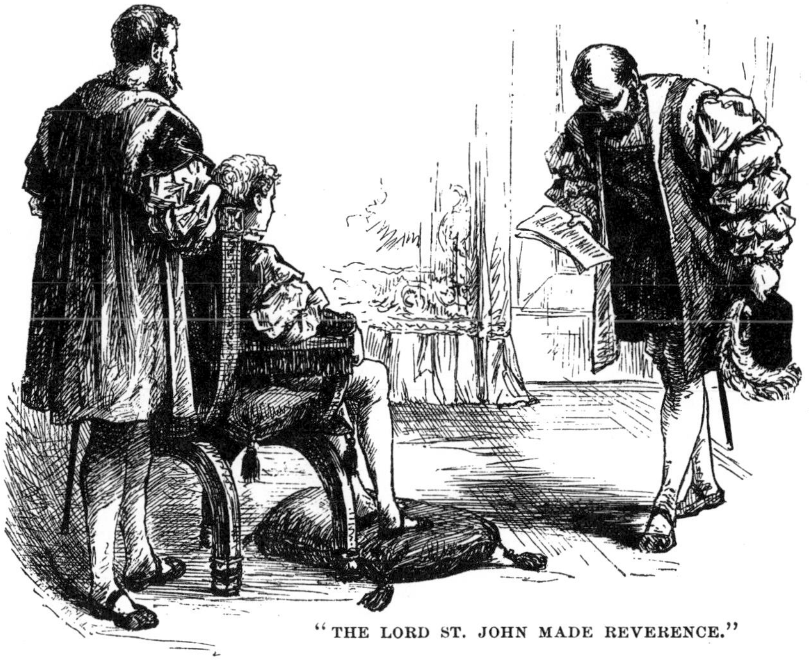

"THE LORD ST. JOHN MADE REVERENCE."

"Thy memory still wrongeth thee, and thou hast shown surprise—but suffer it not to trouble thee, for 'tis a matter that will not bide, but depart with thy mending malady. My lord of Hertford speaketh of the city's banquet which the king's majesty did promise some two months flown, your highness should attend. Thou recallest it now?"

"It grieves me to confess it had indeed escaped me," said Tom, in a hesitating voice; and blushed again.

At this moment the lady Elizabeth and the lady Jane Grey were announced. The two lords exchanged significant glances, and Hertford stepped quickly toward the door. As the young girls passed him, he said in a low voice —

"I pray ye, ladies, seem not to observe his humors, nor show surprise when his memory doth lapse — it will grieve you to note how it doth stick at every trifle."

Meantime lord St. John was saying in Tom's ear —

"Please you sir, keep diligently in mind his majesty's desire. Remember all thou canst — *seem* to remember all else. Let them not perceive that thou art much changed from thy wont, for thou knowest how tenderly thy old play-fellows bear thee in their hearts and how 'twould grieve them. Art willing, sir, that I remain? — and thine uncle?"

Tom signified assent with a gesture and a murmured word, for he was already learning, and in his simple heart was resolved to acquit himself as best he might, according to the king's command.

In spite of every precaution, the conversation among the young people became a little embarrassing, at times. More than once, in truth, Tom was near to breaking down and confessing himself unequal to his tremendous part; but the tact of the princess Elizabeth saved him, or a word from one or the other of the vigilant lords, thrown in apparently by chance, had the same happy effect. Once the little lady Jane turned to Tom and dismayed him with this question, —

"Hast paid thy duty to the queen's majesty to-day, my lord?"

Tom hesitated, looked distressed, and was about to stammer out something at hazard, when lord St. John took the word and answered for him with the easy grace of a courtier accustomed to encounter delicate difficulties and to be ready for them —

"He hath indeed, madam, and she did greatly hearten him, as touching his majesty's condition; is it not so, your highness?"

Tom mumbled something that stood for assent, but felt that he was getting upon dangerous ground. Somewhat later it was mentioned that Tom was to study no more at present, whereupon her little ladyship exclaimed—

"'Tis a pity, 'tis such a pity! Thou wert proceeding bravely. But bide thy time in patience; it will not be for long. Thou'lt yet

HERTFORD AND THE PRINCESSES.

be graced with learning like thy father, and make thy tongue master of as many languages as his, good my prince."

"My father!" cried Tom, off his guard for the moment. "I trow he cannot speak his own so that any but the swine that wallow in the styes may tell his meaning; and as for learning of any sort soever" —

He looked up and encountered a solemn warning in my lord St. John's eyes.

He stopped, blushed, then continued low and sadly: "Ah, my malady persecuteth me again, and my mind wandereth. I meant the king's grace no irreverence."

"We know it, sir," said the princess Elizabeth, taking her "brother's" hand between her two palms, respectfully but caressingly; "trouble not thyself as to that. The fault is none of thine, but thy distemper's."

"Thou'rt a gentle comforter, sweet lady," said Tom, gratefully, "and my heart moveth me to thank thee for't, an' I may be so bold."

Once the giddy little lady Jane fired a simple Greek phrase at Tom. The princess Elizabeth's quick eye saw by the serene blankness of the target's front that the shaft was overshot, so she tranquilly delivered a return volley of sounding Greek on Tom's behalf, and then straightway changed the talk to other matters.

Time wore on pleasantly, and likewise smoothly, on the whole. Snags and sandbars grew less and less frequent, and Tom grew more and more at his ease, seeing that all were so lovingly bent upon helping him and overlooking his mistakes. When it came out that the little ladies were to accompany him to the Lord Mayor's banquet in the evening, his heart gave a bound of relief and delight, for he felt that he should not be friendless, now, among that multitude of strangers, whereas, an hour earlier, the idea of their going with him would have been an insupportable terror to him.

Tom's guardian angels, the two lords, had had less comfort in the interview than the other parties to it. They felt much as if they were piloting a great ship through a dangerous channel; they were on the alert constantly, and found their office no child's play. Wherefore, at last, when the ladies' visit was drawing to a close and the lord Guilford Dudley was announced, they not only felt that their charge had been sufficiently taxed for the present, but also that they themselves were not in the best condition to take their ship back and make their anxious voyage all over again. So they respectfully advised Tom to excuse himself, which he was very glad to do, although a slight shade of disappointment might have been observed upon my lady Jane's face when she heard the splendid stripling denied admittance.

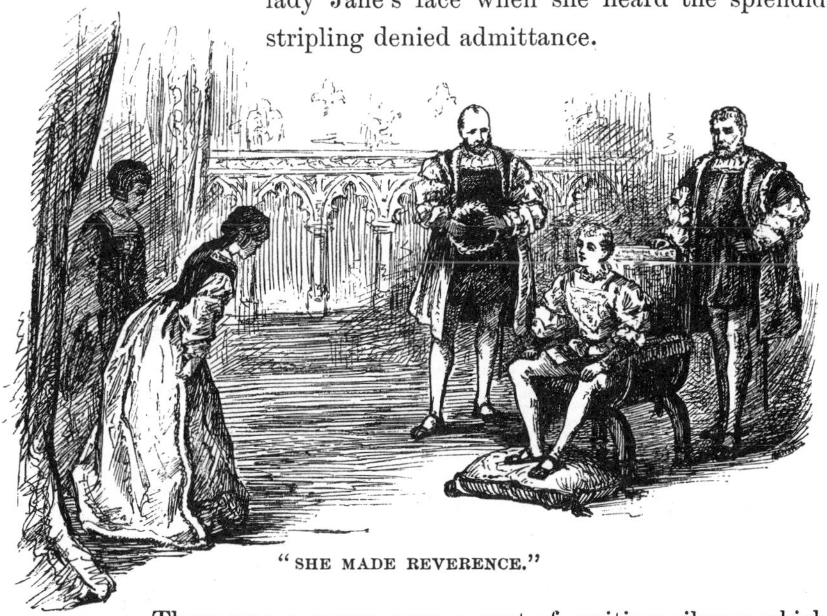

"SHE MADE REVERENCE."

There was a pause, now, a sort of waiting silence which Tom could not understand. He glanced at lord Hertford, who gave him a sign — but he failed to understand that, also. The ready Elizabeth came to the rescue with her usual easy grace. She made reverence and said, —

"Have we leave of the prince's grace my brother to go?"

Tom said —

"Indeed your ladyships can have whatsoever of me they will, for the asking; yet would I rather give them any other thing that in my poor power lieth, than leave to take the light and blessing of their presence hence. Give ye good den, and God be with ye!" Then he smiled inwardly at the thought, "'tis not for nought I have dwelt but among princes in my reading, and taught my tongue some slight trick of their broidered and gracious speech withal!"

When the illustrious maidens were gone, Tom turned wearily to his keepers and said—

"May it please your lordships to grant me leave to go into some corner and rest me?"

Lord Hertford said—

"So please your highness, it is for you to command, it is for us to obey. That thou shouldst rest, is

"OFFERED IT TO HIM ON A GOLDEN SALVER."

indeed a needful thing, since thou must journey to the city presently."

He touched a bell, and a page appeared, who was ordered to desire the presence of Sir William Herbert. This gentleman came straightway, and conducted Tom to an inner apartment. Tom's first movement, there, was to reach for a cup of water; but a silk-and-velvet servitor seized it, dropped upon one knee, and offered it to him on a golden salver.

Next, the tired captive sat down and was going to take off his buskins, timidly asking leave with his eye, but another silk-and-velvet discomforter went down upon his knees and took the office from him. He made two or three further efforts to help himself, but being promptly forestalled each time, he finally gave up, with a sigh of resignation and a murmured "Beshrew me but I marvel they do not require to breathe for me also!" Slippered, and wrapped in a sumptuous robe, he laid himself down at last to rest, but not to sleep, for his head was too full of thoughts and the room too full of people. He could not dismiss the former, so they staid; he did not know enough to dismiss the latter, so they staid also, to his vast regret,—and theirs.

Tom's departure had left his two noble guardians alone. They mused a while, with much head-shaking and walking the floor, then lord St. John said—

"Plainly, what dost thou think?"

"Plainly, then, this. The king is near his end, my nephew is mad, mad will mount the throne, and mad remain. God protect England, since she will need it!"

"Verily it promiseth so, indeed. But . . . have you no misgivings as to . . . as to" . . .

The speaker hesitated, and finally stopped. He evidently felt that he was upon delicate ground. Lord Hertford stopped before him, looked into his face with a clear, frank eye, and said—

"Speak on—there is none to hear but me. Misgivings as to what?"

"I am full loath to word the thing that is in my mind, and thou so near to him in blood, my lord. But craving pardon if I do offend, seemeth it not strange that madness could so change his port and manner! — not but that his port and speech are princely still, but that they *differ* in one unweighty trifle or another, from what his custom

"THEY MUSED A WHILE."

was aforetime. Seemeth it not strange that madness should filch from his memory his father's very lineaments; the customs and observances that are his due from such as be about him; and, leaving him his Latin, strip him of his Greek and French? My lord, be not offended, but ease my mind of its disquiet and receive my grateful thanks. It haunteth me, his saying he was not the prince, and so" —

"Peace, my lord, thou utterest treason! Hast forgot the king's command? Remember I am party to thy crime, if I but listen."

St. John paled, and hastened to say —

"I was in fault, I do confess it. Betray me not, grant me this grace out of thy courtesy, and I will neither think nor speak of this thing more. Deal not hardly with me, sir, else am I ruined."

"I am content, my lord. So thou offend not again, here or in the ears of others, it shall be as though thou hadst not spoken. But thou

"PEACE, MY LORD, THOU UTTEREST TREASON!"

needst not have misgivings. He is my sister's son; are not his voice, his face, his form, familiar to me from his cradle? Madness can do all the odd conflicting things thou seest in him, and more. Dost not recall how that the old Baron Marley, being mad, forgot the favor of his own countenance that he had known for sixty years, and held it was another's; nay, even claimed he was the son of Mary Magdalene, and that his head was made of Spanish glass; and sooth to say, he suffered none to touch it, lest by mischance some heedless hand might shiver it. Give thy misgivings easement, good my lord. This is the very prince, I know him well—and soon will be thy king; it may advantage thee to bear this in mind and more dwell upon it than the other."

"HE BEGAN TO PACE THE FLOOR."

After some further talk, in which the lord St. John covered up his mistake as well as he could by repeated protests that his faith was thoroughly grounded, now, and could not be assailed by doubts again, the lord Hertford relieved his fellow keeper, and sat down to keep watch and ward alone. He was soon deep in meditation. And evidently the longer he thought, the more he was bothered. By and by he began to pace the floor and mutter.

"Tush, he *must* be the prince! Will any he in all the land maintain there can be two, not of one blood and birth, so marvellously twinned? And even were it so, 'twere yet a stranger miracle that chance should cast the one into the other's place. Nay, 'tis folly, folly, folly!"

Presently he said —

"Now were he impostor and called himself prince, look you *that* would be natural; that would be reasonable. But lived ever an impostor yet, who, being called prince by the king, prince by the court, prince by all, *denied* his dignity and pleaded against his exaltation? *No!* By the soul of St. Swithin, no! This is the true prince, gone mad!"

Tom's
FIRST ROYAL DINNER

CHAPTER VII.

TOM'S FIRST ROYAL DINNER.

Somewhat after one in the afternoon, Tom resignedly underwent the ordeal of being dressed for dinner. He found himself as finely clothed as before, but every thing different, every thing changed, from his ruff to his stockings. He was presently conducted with much state to a spacious and ornate apartment, where a table was already set for one. Its furniture was all of massy gold, and beautified with designs which well-nigh made it priceless, since they were the work of Benvenuto. The room was half filled with noble servitors. A chaplain said grace, and Tom was about to fall to, for hunger had long been constitutional with him, but was interrupted by my lord the Earl of Berkeley, who fastened a napkin about his neck; for the great post of Diaperers to the Princes of Wales was hereditary in this nobleman's family. Tom's cup-bearer was present, and forestalled all his attempts to help himself to wine. The Taster to his highness the

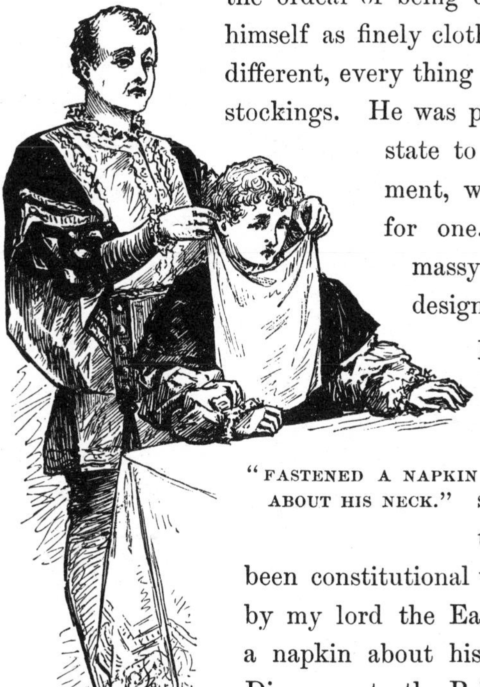

"FASTENED A NAPKIN ABOUT HIS NECK."

Prince of Wales was there also, prepared to taste any suspicious dish upon requirement, and run the risk of being poisoned. He was only an ornamental appendage at this time, and was seldom called upon to exercise his function; but there had been times, not many generations past, when the office of taster had its perils, and was not a grandeur to be desired. Why they did not use a dog or a plumber seems strange; but all the ways of royalty are strange. My lord d'Arcy, First Groom of the Chamber, was there, to do goodness knows what; but there he was — let that suffice. The Lord Chief Butler was there, and stood behind Tom's chair, overseeing the solemnities, under command of the Lord Great Steward and the Lord Head Cook, who stood near. Tom had three hundred and eighty-four servants beside these; but they were not all in that room, of course, nor the quarter of them; neither was Tom aware yet that they existed.

All those that were present had been well drilled within the hour to remember that the prince was temporarily out of his head, and to be careful to show no surprise at his vagaries. These "vagaries" were soon on exhibition before them; but they only moved their compassion and their sorrow, not their mirth. It was a heavy affliction to them to see the beloved prince so stricken.

Poor Tom ate with his fingers mainly; but no one smiled at it, or even seemed to observe it. He inspected his napkin curiously, and with deep interest, for it was of a very dainty and beautiful fabric, then said with simplicity, —

"Prithee take it away, lest in mine unheedfulness it be soiled."

The Hereditary Diaperer took it away with reverent manner, and without word or protest of any sort.

Tom examined the turnips and the lettuce with interest, and asked what they were, and if they were to be eaten; for it was only recently that men had begun to raise these things in England in place of importing them as luxuries from Holland.[1] His question was answered with grave respect, and no surprise manifested. When he had finished his dessert, he filled his pockets with nuts; but nobody appeared to be aware of it, or disturbed by it. But the next moment he was himself disturbed by it, and showed discomposure; for this was the only service he had been permitted to do with his own hands during the meal, and he did not doubt that he had done a most improper and unprincely thing. At that moment the muscles of his nose began to twitch, and the end of that organ to lift and wrinkle. This continued, and Tom began to evince a growing distress. He looked appealingly, first at one and then another of the lords about him, and tears came into his eyes. They sprang forward with dismay in their faces, and begged to know his trouble. Tom said with genuine anguish, —

"TOM ATE WITH HIS FINGERS."

"I crave your indulgence: my nose itcheth cruelly. What is the custom and usage in this emergence? Prithee speed, for 'tis but a little time that I can bear it."

None smiled; but all were sore perplexed, and looked one to the other in deep tribulation for counsel. But behold, here was a dead

[1] See note 4, at end of volume.

wall, and nothing in English history to tell how to get over it. The Master of Ceremonies was not present: there was no one who felt safe to venture upon this uncharted sea, or risk the attempt to solve this solemn problem. Alas! there was no Hereditary Scratcher. Meantime the tears had overflowed their banks, and begun to trickle down Tom's cheeks. His twitching nose was pleading more urgently than ever for relief. At last nature broke down the barriers of etiquette: Tom lifted up an inward prayer for pardon if he was doing wrong, and brought relief to the burdened hearts of his court by scratching his nose himself.

"HE GRAVELY TOOK A DRAUGHT."

His meal being ended, a lord came and held before him a broad, shallow, golden dish with fragrant rose-water in it, to cleanse his mouth and fingers with; and my lord the Hereditary Diaperer stood by with a napkin for his use. Tom gazed at the dish a puzzled moment or two, then raised it to his lips, and gravely took a draught. Then he returned it to the waiting lord, and said, —

"Nay, it likes me not, my lord: it hath a pretty flavor, but it wanteth strength."

This new eccentricity of the prince's ruined mind made all the hearts about him ache; but the sad sight moved none to merriment.

Tom's next unconscious blunder was to get up and leave the table just when the chaplain had taken his stand behind his chair, and with uplifted hands, and closed, uplifted eyes, was in the act of beginning the blessing. Still nobody seemed to perceive that the prince had done a thing unusual.

By his own request, our small friend was now conducted to his private cabinet, and left there alone to his own devices. Hanging upon hooks in the oaken wainscoting were the several pieces of a suit of shining steel armor, covered all over with beautiful designs exquisitely inlaid in gold. This martial panoply belonged to the true prince, — a recent present from Madam Parr the Queen. Tom put on the greaves, the gaunt-

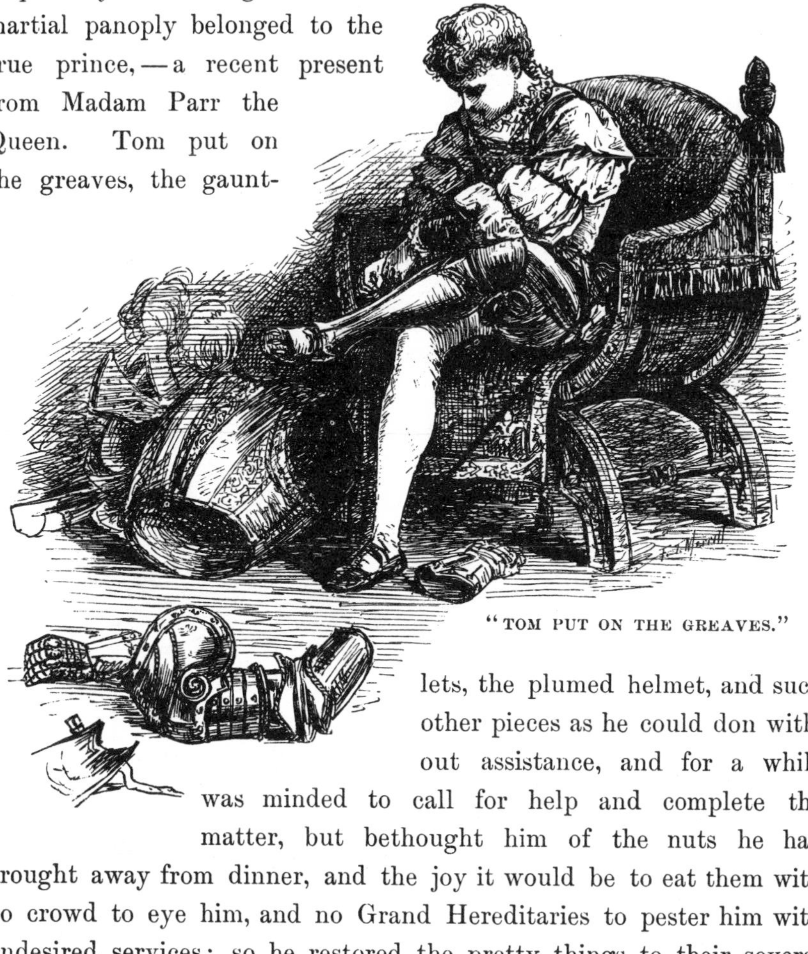

"TOM PUT ON THE GREAVES."

lets, the plumed helmet, and such other pieces as he could don without assistance, and for a while was minded to call for help and complete the matter, but bethought him of the nuts he had brought away from dinner, and the joy it would be to eat them with no crowd to eye him, and no Grand Hereditaries to pester him with undesired services; so he restored the pretty things to their several places, and soon was cracking nuts, and feeling almost naturally happy for the first time since God for his sins had made him a prince. When the nuts were all gone, he stumbled upon some inviting books in a closet, among them one about the etiquette of the English court. This was a prize. He lay down upon a sumptuous divan, and proceeded to instruct himself with honest zeal. Let us leave him there for the present.

The Question of the Seal

CHAPTER VIII.

THE QUESTION OF THE SEAL.

ABOUT five o'clock Henry VIII. awoke out of an unrefreshing nap, and muttered to himself, "Troublous dreams, troublous dreams! Mine end is now at hand: so say these warnings, and my failing pulses do confirm it." Presently a wicked light flamed up in his eye, and he muttered, "Yet will not I die till *he* go before."

His attendants perceiving that he was awake, one of them asked his pleasure concerning the Lord Chancellor, who was waiting without.

"Admit him, admit him!" exclaimed the King eagerly.

The Lord Chancellor entered, and knelt by the King's couch, saying, —

"I have given order, and, according to the King's command, the peers of the realm, in their robes, do now stand at the bar of the House, where, having confirmed the Duke of Norfolk's doom, they humbly wait his majesty's further pleasure in the matter."

The King's face lit up with a fierce joy. Said he, —

"Lift me up! In mine own person will I go before my Parliament, and with mine own hand will I seal the warrant that rids me of" —

His voice failed; an ashen pallor swept the flush from his cheeks; and the attendants eased him back upon his pillows, and hurriedly assisted him with restoratives. Presently he said sorrowfully, —

"Alack, how have I longed for this sweet hour! and lo, too late it cometh, and I am robbed of this so coveted chance. But speed ye, speed ye! let others do this happy office sith 'tis denied to me. I put my great seal in commission: choose thou the lords that shall compose it, and get ye to your work. Speed ye, man! Before the sun shall rise and set again, bring me his head that I may see it."

"According to the King's command, so shall it be. Will't please your majesty to order that the Seal be now restored to me, so that I may forth upon the business?"

"The Seal? Who keepeth the Seal but thou?"

"Please your majesty, you did take it from me two days since,

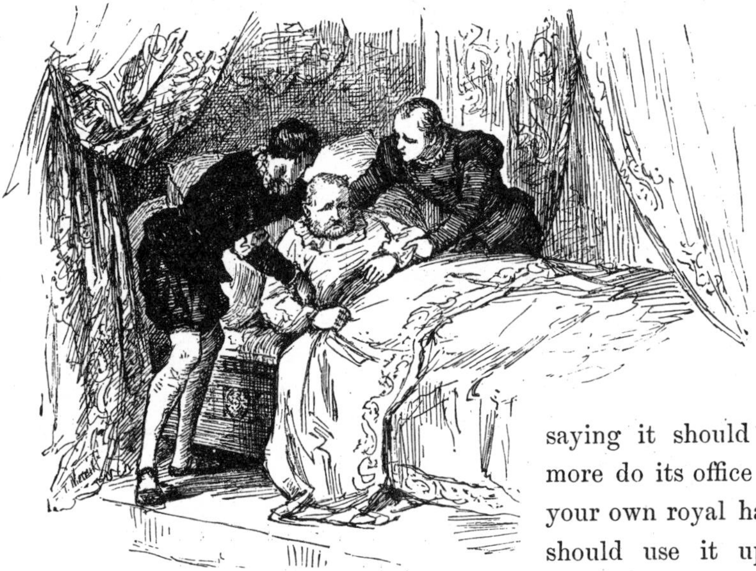

"THE ATTENDANTS EASED HIM BACK UPON HIS PILLOWS."

saying it should no more do its office till your own royal hand should use it upon the Duke of Norfolk's warrant."

"Why, so in sooth I did. I do remember it. . . . What did I with it? . . . I am very feeble. . . . So oft these days doth my memory play the traitor with me. . . . 'Tis strange, strange"—

The King dropped into inarticulate mumblings, shaking his gray head weakly from time to time, and gropingly trying to recollect what he had done with the Seal. At last my lord Hertford ventured to kneel and offer information, —

"Sire, if that I may be so bold, here be several that do remember with me how that you gave the Great Seal into the hands of his highness the Prince of Wales to keep against the day that" —

"True, most true!" interrupted the King. "Fetch it! Go: time flieth!"

Lord Hertford flew to Tom, but returned to the King before very long, troubled and empty-handed. He delivered himself to this effect, —

"It grieveth me, my lord the King, to bear so heavy and unwelcome tidings; but it is the will of God that the prince's affliction abideth still, and he cannot recall to mind that he received the Seal. So came I quickly to report, thinking it were waste of precious time, and little worth withal, that any should attempt to search the long array of chambers and saloons that belong unto his royal high" —

A groan from the King interrupted my lord at this point. After a little while his majesty said, with a deep sadness in his tone, —

"Trouble him no more, poor child. The hand of God lieth heavy upon him, and my heart goeth out in loving compassion for him, and sorrow that I may not bear his burden on mine own old trouble-weighted shoulders, and so bring him peace."

He closed his eyes, fell to mumbling, and presently was silent. After a time he opened his eyes again, and gazed vacantly around until his glance rested upon the kneeling Lord Chancellor. Instantly his face flushed with wrath, —

"What, thou here yet! By the glory of God, an' thou gettest not about that traitor's business, thy mitre shall have holiday the morrow for lack of a head to grace withal!"

The trembling Chancellor answered, —

"Good your majesty, I cry you mercy! I but waited for the Seal."

"Man, hast lost thy wits? The small Seal which aforetime I was wont to take with me abroad lieth in my treasury. And, since the Great Seal hath flown away, shall not it suffice? Hast lost thy wits? Begone! And hark ye, — come no more till thou do bring his head."

The poor Chancellor was not long in removing himself from this dangerous vicinity; nor did the commission waste time in giving the royal assent to the work of the slavish Parliament, and appointing the morrow for the beheading of the premier peer of England, the luckless Duke of Norfolk.[1]

[1] See note 5, at end of volume.

THE RIVER PAGEANT

CHAPTER IX.

THE RIVER PAGEANT.

At nine in the evening the whole vast river-front of the palace was blazing with light. The river itself, as far as the eye could reach citywards, was so thickly covered with watermen's boats and with pleasure-barges, all fringed with colored lanterns, and gently agitated by the waves, that it resembled a glowing and limitless garden of flowers stirred to soft motion by summer winds. The grand terrace of stone steps leading down to the water, spacious enough to mass the army of a German principality upon, was a picture to see, with its ranks of royal halberdiers in polished armor, and its troops of brilliantly costumed servitors flitting up and down, and to and fro, in the hurry of preparation.

Presently a command was given, and immediately all living creatures vanished from the steps. Now the air was heavy with the hush of suspense and expectancy. As far as one's vision could carry, he might see the myriads of people in the boats rise up, and shade their eyes from the glare of lanterns and torches, and gaze toward the palace.

A file of forty or fifty state barges drew up to the steps. They were richly gilt, and their lofty prows and sterns were elaborately carved. Some of them were decorated with banners and streamers; some with cloth-of-gold and arras embroidered with coats-of-arms; others with silken flags that had numberless little silver bells fastened to them, which shook out tiny showers of joyous music whenever the breezes fluttered them; others of yet higher pretensions, sincc they belonged to nobles in the prince's immediate service, had their sides picturesquely fenced with shields gorgeously emblazoned with armorial bearings. Each state barge was towed by a tender. Besides the rowers, these tenders carried each a number of men-at-arms in glossy helmet and breastplate, and a company of musicians.

"A TROOP OF HALBERDIERS APPEARED IN THE GATEWAY."

The advance-guard of the expected procession now appeared in the great gateway, a troop of halberdiers. "They were dressed in striped hose of black and tawny, velvet caps graced at the sides with silver roses, and doublets of murrey and blue cloth, embroidered on the front and back with the three feathers, the prince's blazon, woven in gold. Their halberd staves were covered with crimson velvet, fastened with gilt nails, and ornamented with gold tassels. Filing off on the right and left, they formed two long lines, extending from the gateway of the palace to the water's edge. A thick, rayed cloth or carpet was then unfolded, and laid down between them by attendants in the gold-and-crimson liveries of the prince. This done, a flourish of trumpets resounded from within. A lively prelude arose from the musicians on the water; and two ushers with white wands marched with a slow and stately pace from the portal. They were followed by an officer bearing the civic mace, after whom came another carrying the city's sword; then several sergeants of the city guard, in their full accoutrements, and with badges on their sleeves; then the garter king-at-arms, in his tabard; then several knights of the bath, each with a white lace on his sleeve; then their esquires; then the judges, in their robes of scarlet and coifs; then the lord high chancellor of England, in a robe of scarlet, open before, and purfled with minever; then a deputation of aldermen, in their scarlet cloaks; and then the heads of the different civic companies, in their robes of state. Now came twelve French gentlemen, in splendid habiliments, consisting of pourpoints of white damask barred with gold, short mantles of crimson velvet lined with violet taffeta, and carnation-colored *hauts-de-chausses*, and took their way down the steps. They were of the suite of the French ambassador, and were followed by twelve cavaliers of the suite of the Spanish ambassador, clothed in black velvet, unrelieved by any ornament. Following these came several great English nobles with their attendants."

There was a flourish of trumpets within; and the prince's uncle, the future great Duke of Somerset, emerged from the gateway, arrayed in a "doublet of black cloth-of-gold, and a cloak of crimson satin flowered with gold, and ribanded with nets of silver." He turned, doffed his plumed cap, bent his body in a low reverence, and began to step backward, bowing at each step. A prolonged trumpet-blast followed, and a proclamation, "Way for the high and mighty, the Lord Edward, Prince of Wales!" High aloft on the palace walls a long line of red tongues of flame leaped forth with a thunder-crash: the massed world on the river burst into a mighty roar of welcome; and Tom Canty, the cause and hero of it all, stepped into view, and slightly bowed his princely head.

"TOM CANTY STEPPED INTO VIEW."

He was "magnificently habited in a doublet of white satin, with a front-piece of purple cloth-of-tissue, powdered with diamonds, and edged with ermine. Over this he wore a mantle of white cloth-of-gold, pounced with the triple-feather crest, lined with blue satin, set with pearls and precious stones, and fastened with a clasp of brilliants. About his neck hung the order of the Garter, and several princely foreign orders;" and wherever light fell upon him jewels responded with a blinding flash. O Tom Canty, born in a hovel, bred in the gutters of London, familiar with rags and dirt and misery, what a spectacle is this!

The Prince in the Toils

CHAPTER X.

THE PRINCE IN THE TOILS.

WE left John Canty dragging the rightful prince into Offal Court, with a noisy and delighted mob at his heels. There was but one person in it who offered a pleading word for the captive, and he was not heeded: he was hardly even heard, so great was the turmoil. The prince continued to struggle for freedom, and to rage against the treatment he was suffering, until John Canty lost what little patience was left in him, and raised his oaken cudgel in a sudden fury over the prince's head. The single pleader for the lad sprang to stop the man's arm, and the blow descended upon his own wrist. Canty roared out, —

"Thou'lt meddle, wilt thou? Then have thy reward."

His cudgel crashed down upon the meddler's head: there was a groan, a dim form sank to the ground among the feet of the crowd, and the next moment it lay there in the dark alone. The mob pressed on, their enjoyment nothing disturbed by this episode.

Presently the prince found himself in John Canty's abode, with the door closed against the outsiders. By the vague light of a tallow candle which was thrust into a bottle, he made out the main features of the loathsome den, and also the occupants of it. Two frowsy girls and a middle-aged woman cowered against the wall in one corner, with the aspect of animals habituated to harsh usage, and expecting and dreading it now. From another corner stole a withered hag with streaming gray hair and malignant eyes. John Canty said to this one, —

"Tarry! There's fine mummeries here. Mar them not till thou'st enjoyed them; then let thy hand be heavy as thou wilt. Stand forth, lad. Now say thy foolery again, an' thou'st not forgot it. Name thy name. Who art thou?"

The insulted blood mounted to the little prince's cheek once more, and he lifted a steady and indignant gaze to the man's face, and said,—

"'Tis but ill-breeding in such as thou to command me to speak. I tell thee now, as I told thee before, I am Edward, Prince of Wales, and none other."

"A DIM FORM SANK TO THE GROUND."

The stunning surprise of this reply nailed the hag's feet to the floor where she stood, and almost took her breath. She stared at the prince in stupid amazement, which so amused her ruffianly son, that he burst into a roar of laughter. But the effect upon Tom Canty's mother and sisters was different. Their dread of bodily injury gave way at once to distress of a different sort. They ran forward with woe and dismay in their faces, exclaiming,—

"WHO ART THOU?"

"O poor Tom, poor lad!"

The mother fell on her knees before the prince, put her hands upon his shoulders, and gazed yearningly into his face through her rising tears. Then she said, —

"O my poor boy! thy foolish reading hath wrought its woful work at last, and ta'en thy wit away. Ah! why didst thou cleave to it when I so warned thee 'gainst it? Thou'st broke thy mother's heart."

The prince looked into her face, and said gently, —

"Thy son is well, and hath not lost his wits, good dame. Comfort thee: let me to the palace where he is, and straightway will the King my father restore him to thee."

"The King thy father! O my child! unsay these words that be freighted with death for thee, and ruin for all that be near to thee. Shake off this grewsome dream. Call back thy poor wandering memory. Look upon me. Am not I thy mother that bore thee, and loveth thee?"

The prince shook his head, and reluctantly said,—

"God knoweth I am loath to grieve thy heart; but truly have I never looked upon thy face before."

The woman sank back to a sitting posture on the floor, and, covering her eyes with her hands, gave way to heartbroken sobs and wailings.

"Let the show go on!" shouted Canty. "What, Nan! what, Bet! Mannerless wenches! will ye stand in the prince's presence? Upon your knees, ye pauper scum, and do him reverence!"

He followed this with another horse-laugh. The girls began to plead timidly for their brother; and Nan said,—

"An' thou wilt but let him to bed, father, rest and sleep will heal his madness: prithee, do."

"Do, father," said Bet: "he is more worn than is his wont. To-morrow will he be himself again, and will beg with diligence, and come not empty home again."

This remark sobered the father's joviality, and brought his mind to business. He turned angrily upon the prince, and said,—

"The morrow must we pay two pennies to him that owns this hole; two pennies, mark ye,—all this money for a half-year's rent, else out of this we go. Show what thou'st gathered with thy lazy begging."

The prince said,—

"Offend me not with thy sordid matters. I tell thee again I am the King's son."

A sounding blow upon the prince's shoulder from Canty's broad palm sent him staggering into goodwife Canty's arms, who clasped him to her breast, and sheltered him from a pelting rain of cuffs and slaps by interposing her own person. The frightened girls retreated to their corner; but the grandmother stepped eagerly forward to assist her son. The prince sprang away from Mrs. Canty, exclaiming, —

"Thou shalt not suffer for me, madam. Let these swine do their will upon me alone."

This speech infuriated the swine to such a degree that they set about their work without waste of time. Between them they belabored the boy right soundly, and then gave the girls and their mother a beating for showing sympathy for the victim.

"SENT HIM STAGGERING INTO GOODWIFE CANTY'S ARMS."

"Now," said Canty, "to bed, all of ye. The entertainment has tired me."

The light was put out, and the family retired. As soon as the snorings of the head of the house and his mother showed that they were asleep, the young girls crept to where the prince lay, and covered him tenderly from the cold with straw and rags; and their mother crept to him also, and stroked his hair, and cried over him, whispering broken words of comfort and compassion in his ear the while. She had saved a morsel for him to eat, also; but the boy's pains had swept away all appetite, — at least for black and tasteless crusts. He was touched by her brave and costly defence of him, and by her commiseration; and he thanked her in very noble and princely words, and begged her to go to her sleep and try to forget her sorrows. And he added that the King his father would not let her loyal kindness and devotion go unrewarded. This return to his "madness" broke her heart anew, and she strained him to her breast again and again and then went back, drowned in tears, to her bed.

As she lay thinking and mourning, the suggestion began to creep into her mind that there was an undefinable something about this boy that was lacking in Tom Canty, mad or sane. She could not describe it, she could not tell just what it was, and yet her sharp mother-instinct seemed to detect it and perceive it. What if the boy were really not her son, after all? O, absurd! She almost smiled at the idea, spite of her griefs and troubles. No matter, she found that it was an idea that would not "down," but persisted in haunting her. It pursued her, it harassed her, it clung to her, and refused to be put away or ignored. At last she perceived that there was not going to be any peace for her until she should devise a test that should prove, clearly and without question, whether this lad was her son or not, and so banish these wearing and worrying doubts. Ah yes, this was plainly the right way out of the difficulty; therefore she set her wits to work at once to contrive that test. But it was an easier thing to propose than to accomplish. She turned over in her mind one promising test after another, but was obliged to relinquish them all — none of them were absolutely sure, absolutely perfect; and an imperfect one could not satisfy her. Evidently she was racking her head in vain — it seemed manifest that she must give the matter up. While this depressing thought was passing through her mind, her ear caught the regular breathing of the boy, and she knew he had fallen asleep. And while she listened, the measured breathing was broken by a soft, startled cry, such as one utters in a troubled dream. This chance occurrence furnished her instantly with a plan worth all her labored tests combined. She at once set herself feverishly, but noiselessly, to work, to relight her candle, muttering to herself, "Had I but seen him *then*, I should have known! Since that day, when he was little, that the powder burst in his face, he hath never been startled of a sudden out of his dreams or out of his thinkings, but he hath cast his hand before his eyes, even as he did that day; and not as others would do it, with the palm inward, but always with the palm turned outward — I have seen it a hundred times, and it hath never varied nor ever failed. Yes, I shall soon know, now!"

By this time she had crept to the slumbering boy's side, with the

"SHE BENT HEEDFULLY AND WARILY OVER HIM."

candle, shaded, in her hand. She bent heedfully and warily over him, scarcely breathing, in her suppressed excitement, and suddenly flashed the light in his face and struck the floor by his ear with her knuckles. The sleeper's eyes sprung wide open, and he cast a startled stare about him — but he made no special movement with his hands.

The poor woman was smitten almost helpless with surprise and grief; but she contrived to hide her emotions, and to soothe the boy to sleep again; then she crept apart and communed miserably with herself upon the disastrous result of her experiment. She tried to believe that her Tom's madness had banished this habitual gesture of his; but she could not do it. "No," she said, "his *hands* are not mad, they could not unlearn so old a habit in so brief a time. O, this is a heavy day for me!"

Still, hope was as stubborn, now, as doubt had been before; she could not bring herself to accept the verdict of the test; she must try the thing again — the failure must have been only an accident; so she startled the boy out of his sleep a second and a third time, at intervals — with the same result which had marked the first test — then she dragged herself to bed, and fell sorrowfully asleep, saying, "But I cannot give him up — O, no, I cannot, I cannot — he *must* be my boy!"

"THE PRINCE SPRANG UP."

The poor mother's interruptions having ceased, and the prince's pains having gradually lost their power to disturb him, utter weariness at last sealed his eyes in a profound and restful sleep. Hour after hour slipped away, and still he slept like the dead. Thus four or five hours passed. Then his stupor began to lighten. Presently while half asleep and half awake, he murmured —

"Sir William!"

After a moment —

"Ho, Sir William Herbert! Hie thee hither, and list to the strangest dream that ever . . . Sir William! dost hear? Man, I did think me changed to a pauper, and . . . Ho there! Guards! Sir William! What! is there no groom of the chamber in waiting? Alack it shall go hard with" —

"What aileth thee?" asked a whisper near him. "Who art thou calling?"

"Sir William Herbert. Who art thou?"

"I? Who should I be, but thy sister Nan? O, Tom, I had forgot! Thou'rt mad yet — poor lad thou'rt mad yet, would I had never woke to know it again! But prithee master thy tongue, lest we be all beaten till we die!"

The startled prince sprang partly up, but a sharp reminder from his stiffened bruises brought him to himself, and he sunk back among his foul straw with a moan and the ejaculation —

"Alas, it was no dream, then!"

In a moment all the heavy sorrow and misery which sleep had banished were upon him again, and he realized that he was no longer a petted prince in a palace, with the adoring eyes of a nation upon him, but a pauper, an outcast, clothed in rags, prisoner in a den fit only for beasts, and consorting with beggars and thieves.

In the midst of his grief he began to be conscious of hilarious noises and shoutings, apparently but a block or two away. The next moment there were several sharp raps at the door; John Canty ceased from snoring and said —

"Who knocketh? What wilt thou?"

A voice answered —

"Know'st thou who it was thou laid thy cudgel on?"

"No. Neither know I, nor care."

"Belike thou'lt change thy note eftsoons. An' thou would save thy neck, nothing but flight may stead thee. The man is this moment delivering up the ghost. 'Tis the priest, Father Andrew!"

"God-a-mercy!" exclaimed Canty. He roused his family, and hoarsely commanded, "Up with ye all and fly — or bide where ye are and perish!"

Scarcely five minutes later the Canty household were in the street and flying for their lives. John Canty held the prince by the wrist, and hurried him along the dark way, giving him this caution in a low voice —

"HURRIED HIM ALONG THE DARK WAY."

"Mind thy tongue, thou mad fool, and speak not our name. I will choose me a new name, speedily, to throw the law's dogs off the scent. Mind thy tongue, I tell thee!"

He growled these words to the rest of the family —

"If it so chance that we be separated, let each make for London bridge; whoso findeth himself as far as the last linen-draper's shop on the bridge, let him tarry there till the others be come, then will we flee into Southwark together."

At this moment the party burst suddenly out of darkness into light; and not only into light, but into the midst of a multitude of singing, dancing, and shouting people, massed together on the river frontage. There was a line of bonfires stretching as far as one could see, up and down the Thames; London bridge was illuminated; Southwark bridge likewise; the entire river was aglow with the flash and sheen of colored lights; and constant explosions of fireworks filled the skies with an intricate commingling of shooting splendors and a thick rain of dazzling sparks that almost turned night into day; everywhere were crowds of revellers; all London seemed to be at large.

John Canty delivered himself of a furious curse and commanded a retreat; but it was too late. He and his tribe were swallowed up in that swarming hive of humanity, and hopelessly separated from each other in an instant. We are not considering that the prince was one of his tribe; Canty still kept his grip upon him. The prince's heart was beating high with hopes of escape, now. A burly waterman, considerably exalted with liquor, found himself rudely shoved, by Canty, in his efforts to plough through the crowd; he laid his great hand on Canty's shoulder and said —

"Nay, whither so fast, friend? Dost canker thy soul with sordid business when all that be leal men and true make holiday?"

"Mine affairs are mine own, they concern thee not," answered Canty, roughly; "take away thy hand and let me pass."

"Sith that is thy humor, thou'lt *not* pass, till thou'st drunk to the Prince of Wales, I tell thee that," said the waterman, barring the way resolutely.

"Give me the cup, then, and make speed, make speed!"

Other revellers were interested by this time. They cried out —

"The loving-cup, the loving-cup! make the sour knave drink the loving-cup, else will we feed him to the fishes."

So a huge loving-cup was brought; the waterman, grasping it by one of its handles, and with his other hand bearing up the end of an imaginary napkin, presented it in due and ancient form to Canty, who had to grasp the opposite handle with one of his hands and take off the lid with the other, according to ancient custom.[1] This left the prince hand-free for a second, of course. He wasted no time, but dived among the forest of legs about him and disappeared. In another moment he could not have been harder to find, under that tossing sea of life, if its billows had been the Atlantic's and he a lost sixpence.

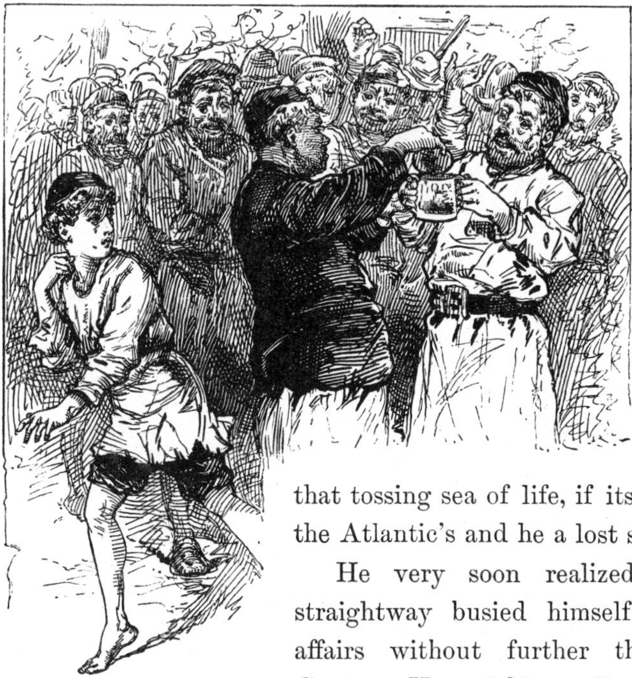

"HE WASTED NO TIME."

He very soon realized this fact, and straightway busied himself about his own affairs without further thought of John Canty. He quickly realized another thing, too. To wit, that a spurious Prince of Wales was being feasted by the city in his stead. He easily concluded that the pauper lad, Tom Canty, had deliberately taken advantage of his stupendous opportunity and become a usurper.

Therefore there was but one course to pursue — find his way to the Guildhall, make himself known, and denounce the impostor. He also made up his mind that Tom should be allowed a reasonable time for spiritual preparation, and then be hanged, drawn and quartered, according to the law and usage of the day, in cases of high treason.

[1] See Note 6, at end of volume.

AT GUILDHALL

CHAPTER XI.

AT GUILDHALL.

THE royal barge, attended by its gorgeous fleet, took its stately way down the Thames through the wilderness of illuminated boats. The air was laden with music; the river banks were beruffled with joy-flames; the distant city lay in a soft luminous glow from its countless invisible bonfires; above it rose many a slender spire into the sky, incrusted with sparkling lights, wherefore in their remoteness they seemed like jewelled lances thrust aloft; as the fleet swept along, it was greeted from the banks with a continuous hoarse roar of cheers and the ceaseless flash and boom of artillery.

To Tom Canty, half buried in his silken cushions, these sounds and this spectacle were a wonder unspeakably sublime and astonishing. To his little friends at his side, the princess Elizabeth and the lady Jane Grey, they were nothing.

Arrived at the Dowgate, the fleet was towed up the limpid Walbrook (whose channel has now been for two centuries buried out of sight under acres of buildings,) to Bucklersbury, past houses and under bridges populous with merry-makers and brilliantly lighted, and at last came to a halt in a basin where now is Barge Yard, in the centre of the ancient city of London. Tom disembarked, and he and his gallant procession crossed Cheapside and made a short march through the Old Jewry and Basinghall street to the Guildhall.

Tom and his little ladies were received with due ceremony by the Lord Mayor and the Fathers of the City, in their gold chains and scarlet robes of state, and conducted to a rich canopy of state at the head of the great hall, preceded by heralds making proclamation, and by the Mace and the City Sword. The lords and ladies who were to attend upon Tom and his two small friends took their places behind their chairs.

At a lower table the court grandees and other guests of noble degree were seated, with the magnates of the city; the commoners took places at a multitude of tables on the main floor of the hall. From their lofty vantage-ground, the giants Gog and Magog, the ancient guardians of the city, contemplated the spectacle below them with eyes

"A RICH CANOPY OF STATE."

grown familiar to it in forgotten generations. There was a bugle-blast and a proclamation, and a fat butler appeared in a high perch in the leftward wall, followed by his servitors bearing with impressive solemnity a royal Baron of Beef, smoking hot and ready for the knife.

After grace, Tom (being instructed) rose — and the whole house with him — and drank from a portly golden loving-cup with the princess Elizabeth; from her it passed to the lady Jane, and then traversed the general assemblage. So the banquet began.

By midnight the revelry was at its height. Now came one of those picturesque spectacles so admired in that old day. A description of it is still extant in the quaint wording of a chronicler who witnessed it:

"Space being made, presently entered a baron and an earl appareled after the Turkish fashion in long robes of bawdkin powdered with gold; hats on their heads of crimson velvet, with great rolls of gold, girded with two swords, called scimitars, hanging by great bawdricks of gold. Next came yet another baron and another earl, in two long gowns of yellow satin, traversed with white satin, and in every bend of white was a bend of crimson satin, after the fashion of Russia, with furred hats of gray on their heads; either of them having an hatchet in their hands, and boots with *pykes*" (points a foot long), "turned up. And after them came a knight, then the Lord High Admiral, and with him five nobles, in doublets of crimson velvet, voyded low on the back and before to the cannell-bone, laced on the breasts with chains of silver; and, over that, short cloaks of crimson satin, and on their heads hats after the dancers' fashion, with pheasants' feathers in them. These were appareled after the fashion of Prussia. The torch-bearers, which were about an hundred, were appareled in crimson satin and green, like Moors, their faces black. Next came in a *mommarye*. Then the minstrels, which were disguised, danced; and the lords and ladies did wildly dance also, that it was a pleasure to behold."

And while Tom, in his high seat, was gazing upon this "wild" dancing, lost in admiration of the dazzling commingling of kaleidoscopic colors which the whirling turmoil of gaudy figures below him presented, the ragged but real little prince of Wales was proclaiming his rights and his wrongs, denouncing the impostor, and clamoring for admission at the gates of Guildhall! The crowd enjoyed this episode prodigiously, and pressed forward and craned their necks to see the small rioter. Presently they began to taunt him and mock at him, purposely to goad him into a higher and still more entertaining fury. Tears of mortification sprung to his eyes, but he stood his ground and defied the mob right royally. Other taunts followed, added mockings stung him, and he exclaimed —

"I tell ye again, you pack of unmannerly curs, I am the prince of Wales! And all forlorn and friendless as I be, with none to give me word of grace or help me in my need, yet will not I be driven from my ground, but will maintain it!"

"Though thou be prince or no prince, 'tis all one, thou be'st a gallant lad, and not friendless neither! Here stand I by thy side to prove it; and mind I tell thee thou might'st have a worser friend than Miles Hendon and yet not tire thy legs with seeking. Rest thy small jaw, my child, I talk the language of these base kennel-rats like to a very native."

The speaker was a sort of Don Cæsar de Bazan in dress, aspect, and bearing. He was tall, trim-built, muscular. His doublet and trunks were of rich material, but faded and threadbare, and their gold-lace adornments were sadly tarnished; his ruff was rumpled and damaged; the plume in his slouched hat was broken and had a bedraggled and disreputable look; at his side he wore a long rapier in a rusty iron sheath; his swaggering carriage marked him at once as a ruffler of the camp. The speech of this fantastic figure was received with an explosion of jeers and laughter. Some cried, "'Tis another prince in disguise!" "'Ware thy tongue, friend, belike he is dangerous!" "Marry, he looketh it — mark his eye!" "Pluck the lad from him — to the horse-pond wi' the cub!"

Instantly a hand was laid upon the prince, under the impulse of this happy thought; as instantly the stranger's long sword was out and the meddler went to the earth under a sounding thump with the flat of it. The next moment a score of voices shouted "Kill the dog! kill him! kill him!" and the mob closed in on the warrior, who

"BEGAN TO LAY ABOUT HIM."

backed himself against a wall and began to lay about him with his long weapon like a madman. His victims sprawled this way and that, but the mob-tide poured over their prostrate forms and dashed itself against the champion with undiminished fury. His moments seemed numbered, his destruction certain, when suddenly a trumpet-blast sounded, a voice shouted, "Way for the king's messenger!" and a troop of horsemen came charging down upon the mob, who fled out of harm's reach as fast as their legs could carry them. The bold stranger caught up the prince in his arms, and was soon far away from danger and the multitude.

Return we within the Guildhall. Suddenly, high above the jubilant roar and thunder of the revel, broke the clear peal of a bugle-note. There was instant silence, — a deep hush; then a single voice rose — that of the messenger from the palace — and began to pipe forth a proclamation, the whole multitude standing, listening. The closing words, solemnly pronounced, were —

"The king is dead!"

The great assemblage

"LONG LIVE THE KING!"

bent their heads upon their breasts with one accord; remained so, in profound silence, a few moments; then all sunk upon their knees in a body, stretched out their hands toward Tom, and a mighty shout burst forth that seemed to shake the building —

"Long live the king!"

Poor Tom's dazed eyes wandered abroad over this stupefying spectacle, and finally rested dreamily upon the kneeling princesses beside him, a moment, then upon the earl of Hertford. A sudden purpose dawned in his face. He said, in a low tone, at lord Hertford's ear —

"Answer me truly, on thy faith and honor! Uttered I here a command, the which none but a king might hold privilege and prerogative to utter, would such commandment be obeyed, and none rise up to say me nay?"

"None, my liege, in all these realms. In thy person bides the majesty of England. Thou art the king — thy word is law."

Tom responded, in a strong, earnest voice, and with great animation —

"Then shall the king's law be law of mercy, from this day, and never more be law of blood! Up from thy knees and away! To the Tower and say the king decrees the duke of Norfolk shall not die!"[1]

The words were caught up and carried eagerly from lip to lip far and wide over the hall, and as Hertford hurried from the presence, another prodigious shout burst forth —

"The reign of blood is ended! Long live Edward, king of England!"

[1] See Note 7, at end of volume.

The Prince and His Deliverer

CHAPTER XII.

THE PRINCE AND HIS DELIVERER.

As soon as Miles Hendon and the little prince were clear of the mob, they struck down through back lanes and alleys toward the river. Their way was unobstructed until they approached London Bridge; then they ploughed into the multitude again, Hendon keeping a fast grip upon the prince's — no, the king's — wrist. The tremendous news was already abroad, and the boy learned it from a thousand voices at once — "The king is dead!" The tidings struck a chill to the heart of the poor little waif, and sent a shudder through his frame. He realized the greatness of his loss, and was filled with a bitter grief; for the grim tyrant who had been such a terror to others had always been gentle with him. The tears sprung to his eyes and blurred all objects. For an instant he felt himself the most forlorn, outcast, and forsaken of God's creatures — then another cry shook the night with its far-reaching thunders: "Long live King Edward the Sixth!" and this made his eyes kindle, and thrilled him with pride to his fingers' ends. "Ah," he thought, "how grand and strange it seems — I AM KING!"

Our friends threaded their way slowly through the throngs upon the Bridge. This structure, which had stood for six hundred years, and had been a noisy and populous thoroughfare all that time, was a curious affair, for a closely packed rank of stores and shops, with family quarters overhead, stretched along both sides of it, from one bank of the river to the other. The Bridge was a sort of town to itself; it had its inn, its beer houses, its bakeries, its haberdasheries, its food markets, its manufacturing industries, and even its church. It looked upon the two neighbors which it linked together — London and Southwark — as being well enough, as suburbs, but not otherwise par-

"OUR FRIENDS THREADED THEIR WAY."

ticularly important. It was a close corporation, so to speak; it was a narrow town, of a single street a fifth of a mile long, its population was but a village population, and everybody in it knew all his fellow townsmen intimately, and had known their fathers and mothers before them — and all their little family affairs into the bargain. It had its aristocracy, of course — its fine old families of butchers, and bakers, and what-not, who had occupied the same old premises for five or six hundred years, and knew the great history of the Bridge from beginning to end, and all its strange legends; and who always talked bridgy talk, and thought bridgy thoughts, and lied in a long, level, direct, substantial bridgy way. It was just the sort of population to be narrow and ignorant and self-conceited. Children were born on the Bridge, were reared there, grew to old age and finally died without ever having set a foot upon any part of the world but London Bridge alone. Such people would naturally imagine that the mighty and interminable procession which moved through its street night and day, with its confused roar of shouts and cries, its neighings and bellowings and bleatings and its muffled thunder-tramp, was the one great thing in this world, and themselves somehow the proprietors of it. And so they were, in effect — at least they could exhibit it from their windows, and did — for a consideration — whenever a returning king or hero gave it a fleeting splendor, for there was no place like it for affording a long, straight, uninterrupted view of marching columns.

Men born and reared upon the Bridge found life unendurably dull and inane, elsewhere. History tells of one of these who left the Bridge at the age of seventy-one and retired to the country. But he could only fret and toss in his bed; he could not go to sleep, the deep stillness was so painful, so awful, so oppressive. When he was worn out with it, at last, he fled back to his old home, a lean and haggard spectre, and fell peacefully to rest and pleasant dreams under the lulling music of the lashing waters and the boom and crash and thunder of London Bridge.

In the times of which we are writing, the Bridge furnished "object lessons" in English history, for its children — namely, the livid and decaying heads of renowned men impaled upon iron spikes atop of its gateways. But we digress.

Hendon's lodgings were in the little inn on the Bridge. As he neared the door with his small friend, a rough voice said —

"So, thou'rt come at last! Thou'lt not escape again, I warrant thee; and if pounding thy bones to a pudding can teach thee somewhat, thou'lt not keep us waiting another time, mayhap" — and John Canty put out his hand to seize the boy.

Miles Hendon stepped in the way and said —

"OBJECT LESSONS" IN ENGLISH HISTORY.

"Not too fast, friend. Thou art needlessly rough, methinks. What is the lad to thee?"

"If it be any business of thine to make and meddle in others' affairs, he is my son."

"'Tis a lie!" cried the little king, hotly.

"Boldly said, and I believe thee, whether thy small head-piece be sound or cracked, my boy. But whether this scurvy ruffian be thy father or no, 'tis all one, he shall not have thee to beat thee and abuse, according to his threat, so thou prefer to bide with me."

"I do, I do — I know him not, I loathe him, and will die before I will go with him."

"Then 'tis settled, and there is nought more to say."

"We will see, as to that!" exclaimed John Canty, striding past Hendon to get at the boy; "by force shall he" —

"If thou do but touch him, thou animated offal, I will spit thee like a goose!" said Hendon, barring the way and laying his hand upon his sword hilt. Canty drew back. "Now mark ye," continued Hendon, "I took this lad under my protection when a mob of such as thou would have mishandled him, mayhap killed him; dost imagine I will desert him now to a worser fate?—for whether thou art his father or no,—and sooth to say, I think it is a lie—a decent swift death were better for such a lad than life in such brute hands as thine. So go

"JOHN CANTY MOVED OFF."

thy ways, and set quick about it, for I like not much bandying of words, being not over-patient in my nature."

John Canty moved off, muttering threats and curses, and was swallowed from sight in the crowd. Hendon ascended three flights of stairs to his room, with his charge, after ordering a meal to be sent thither. It was a poor apartment, with a shabby bed and some odds and ends of old furniture in it, and was vaguely lighted by a couple of sickly candles. The little king dragged himself to the bed and lay down upon it, almost exhausted with hunger and fatigue. He had been on his feet a good part of a day and a night, for it was now two or three o'clock in the morning, and had eaten nothing meantime. He murmured drowsily —

"Prithee call me when the table is spread," and sunk into a deep sleep immediately.

A smile twinkled in Hendon's eye, and he said to himself —

"By the mass, the little beggar takes to one's quarters and usurps one's bed with as natural and easy a grace as if he owned them — with never a by-your-leave or so-please-it-you, or any thing of the sort. In his diseased ravings he called himself the prince of Wales, and bravely doth he keep up the character. Poor little friendless rat, doubtless his mind has been disordered with ill usage. Well, I will be his friend; I have saved him, and it draweth me strongly to him; already I love the bold-tongued little rascal. How soldier-like he faced the smutty rabble and flung back his high defiance! And what a comely, sweet and gentle face he hath, now that sleep hath conjured away its troubles and its griefs. I will teach him, I will cure his malady; yea, I will be his elder brother, and care for him and watch over him; and whoso would shame him or do him hurt, may order his shroud, for though I be burnt for it he shall need it!"

He bent over the boy and contemplated him with kind and pitying interest, tapping the young cheek tenderly and smoothing back the tangled curls with his great brown hand. A slight shiver passed over the boy's form. Hendon muttered —

"See, now, how like a man it was to let him lie here uncovered and fill his body with deadly rheums. Now what shall I do? 'twill wake him to take him up and put him within the bed, and he sorely needeth sleep."

He looked about for extra covering, but finding none, doffed his doublet and wrapped the lad in it, saying, "I am used to nipping air and scant apparel, 'tis little I shall mind the cold" — then walked up and down the room to keep his blood in motion, soliloquizing, as before.

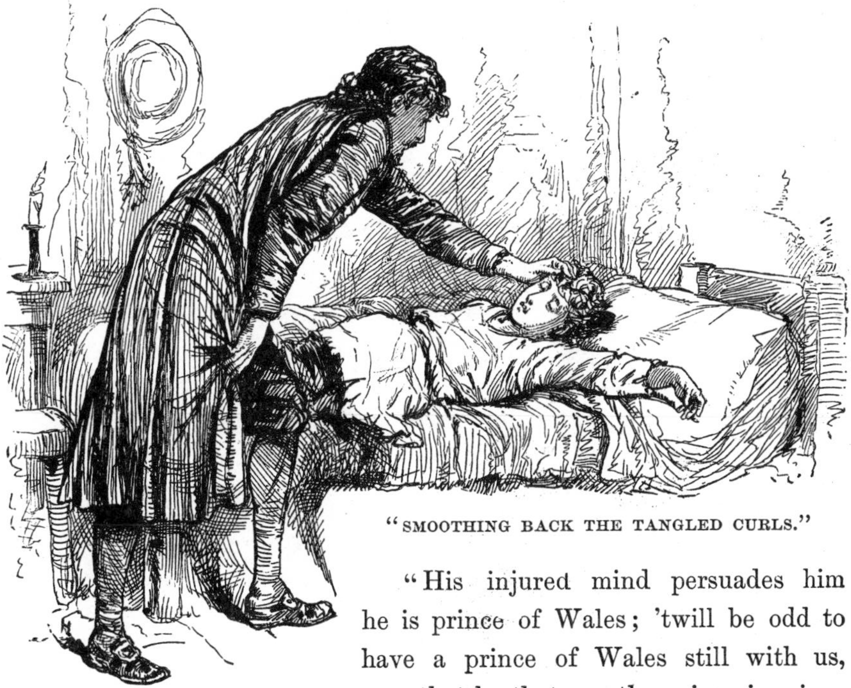

"SMOOTHING BACK THE TANGLED CURLS."

"His injured mind persuades him he is prince of Wales; 'twill be odd to have a prince of Wales still with us, now that he that *was* the prince is prince no more, but king, — for this poor mind is set upon the one fantasy, and will not reason out that now it should cast by the prince and call itself the king. . . . If my father liveth still, after these seven years that I have heard nought from home in my foreign dungeon, he will welcome the poor lad and give him generous shelter for my sake; so will my good elder brother, Arthur; my other brother, Hugh — but I will crack his crown, an' *he* interfere, the fox-hearted, ill-conditioned animal! Yes, thither will we fare — and straightway, too."

A servant entered with a smoking meal, disposed it upon a small deal table, placed the chairs, and took his departure, leaving such cheap lodgers as these to wait upon themselves. The door slammed after him, and the noise woke the boy, who sprung to a sitting posture, and shot a glad glance about him; then a grieved look came into his face and he murmured, to himself, with a deep sigh, "Alack, it was but a dream, woe is me." Next he noticed Miles Hendon's doublet — glanced from that to Hendon, comprehended the sacrifice that had been made for him, and said, gently —

"Thou art good to me, yes, thou art very good to me. Take it and put it on — I shall not need it more."

Then he got up and walked to the washstand in the corner, and stood there, waiting. Hendon said in a cheery voice —

"We'll have a right hearty sup and bite, now, for every thing is savory and smoking hot, and that and thy nap together will make thee a little man again, never fear!"

The boy made no answer, but bent a steady look, that was filled with grave surprise, and also somewhat touched with impatience, upon the tall knight of the sword. Hendon was puzzled, and said —

"What's amiss?"

"Good sir, I would wash me."

"O, is that all! Ask no permission of Miles Hendon for aught thou cravest. Make thyself perfectly free here, and welcome, with all that are his belongings."

Still the boy stood, and moved not; more, he tapped the floor once or twice with his small impatient foot. Hendon was wholly perplexed. Said he —

"Bless us, what is it?"

"Prithee pour the water, and make not so many words!"

Hendon, suppressing a horse-laugh, and saying to himself, "By all the saints, but this is admirable!" stepped briskly forward and did the small insolent's bidding; then stood by, in a sort of stupefaction, until the command, "Come — the towel!" woke him sharply up. He took up a towel, from under the boy's nose, and handed it to him, without comment. He now proceeded to comfort his own face with a wash, and while he was at it his adopted child seated himself at the table and prepared to fall to. Hendon despatched his ablutions with alacrity, then drew back the other chair and was about to place himself at table, when the boy said, indignantly —

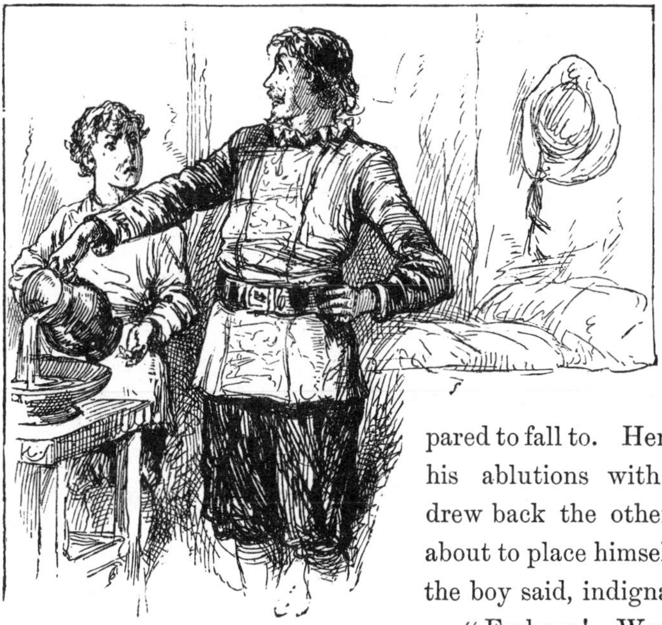

"PRITHEE, POUR THE WATER."

"Forbear! Wouldst sit in the presence of the king?"

This blow staggered Hendon to his foundations. He muttered to himself, "Lo, the poor thing's madness is up with the time! it hath changed with the great change that is come to the realm, and now in fancy is he *king!* Good lack, I must humor the conceit, too — there is no other way — faith, he would order me to the Tower, else!"

And pleased with this jest, he removed the chair from the table, took his stand behind the king, and proceeded to wait upon him in the courtliest way he was capable of.

While the king ate, the rigor of his royal dignity relaxed a little, and with his growing contentment came a desire to talk. He said —

"GO ON — TELL ME THY STORY."

"I think thou callest thyself Miles Hendon, if I heard thee aright?"

"Yes, sire," Miles replied; then observed to himself, "If I *must* humor the poor lad's madness, I must sire him, I must majesty him, I must not go by halves, I must stick at nothing that belongeth to the part I play, else shall I play it ill and work evil to this charitable and kindly cause."

The king warmed his heart with a second glass of wine, and said — "I would know thee — tell me thy story. Thou hast a gallant way with thee, and a noble — art nobly born?"

"We are of the tail of the nobility, good your majesty. My father is a baronet — one of the smaller lords, by knight service[1] — Sir Richard Hendon, of Hendon Hall, by Monk's Holm in Kent."

"The name has escaped my memory. Go on — tell me thy story."

"'Tis not much, your majesty, yet perchance it may beguile a short half hour for want of a better. My father, Sir Richard, is very rich,

[1] He refers to the order of baronets, or baronettes, — the *barones minores*, as distinct from the parliamentary barons; — not, it need hardly be said, the baronets of later creation.

and of a most generous nature. My mother died whilst I was yet a boy. I have two brothers: Arthur, my elder, with a soul like to his father's; and Hugh, younger than I, a mean spirit, covetous, treacherous, vicious, underhanded — a reptile. Such was he from the cradle; such was he ten years past, when I last saw him — a ripe rascal at nineteen, I being twenty, then, and Arthur twenty-two. There is none other of us but the lady Edith, my cousin — she was sixteen, then — beautiful, gentle, good, the daughter of an earl, the last of her race, heiress of a great fortune and a lapsed title. My father was her guardian. I loved her and she loved me; but she was betrothed to Arthur from the cradle, and Sir Richard would not suffer the contract to be broken. Arthur loved another maid, and bade us be of good cheer and hold fast to the hope that delay and luck together would some day give success to our several causes. Hugh loved the lady Edith's fortune, though in truth he said it was herself he loved — but then 'twas his way, alway, to say the one thing and mean the other. But he lost his arts upon the girl; he could deceive my father, but none else. My father loved him best of us all, and trusted and believed him; for he was the youngest child and others hated him — these qualities being in all ages sufficient to win a parent's dearest love; and he had a smooth persuasive tongue, with an admirable gift of lying — and these be qualities which do mightily assist a blind affection to cozen itself. I was wild — in troth I might go yet farther and say *very* wild, though 'twas a wildness of an innocent sort, since it hurt none but me, brought shame to none, nor loss, nor had in it any taint of crime or baseness, or what might not beseem mine honorable degree.

"Yet did my brother Hugh turn these faults to good account — he seeing that our brother Arthur's health was but indifferent, and hoping the worst might work him profit were I swept out of the path — so, — but 'twere a long tale, good my liege, and little worth the telling. Briefly, then, this brother did deftly magnify my faults and make them crimes; ending his base work with finding a silken ladder in mine apartments — conveyed thither by his own means — and did convince my father by this, and suborned evidence of servants and other lying knaves, that I was minded to carry off my Edith and marry with her, in rank defiance of his will.

"Three years of banishment from home and England might make a soldier and a man of me, my father said, and teach me some degree of wisdom. I fought out my long probation in the continental wars, tasting sumptuously of hard knocks, privation and adventure; but in my last battle I was taken captive, and during the seven years that have waxed and waned since then, a foreign dungeon hath harbored me. Through wit and courage I won to the free air at last, and fled hither straight; and am but just arrived, right poor in purse and raiment, and poorer still in knowledge of what these dull seven years have wrought at Hendon Hall, its people and belongings. So please you, sir, my meagre tale is told."

"Thou hast been shamefully abused!" said the little king, with a flashing eye. "But I will right thee — by the cross will I! The king hath said it."

Then, fired by the story of Miles's wrongs, he loosed his tongue and poured the history of his own recent misfortunes into the ears of his astonished listener. When he had finished, Miles said to himself —

"Lo, what an imagination he hath! Verily this is no common mind; else, crazed or sane, it could not weave so straight and gaudy a tale as this out of the airy nothings wherewith it hath wrought this curious romaunt. Poor ruined little head, it shall not lack friend or shelter whilst I bide with the living. He shall never leave my side; he shall be my pet, my little comrade. And he shall be cured! — aye, made whole and sound — then will he make himself a name — and proud shall I be to say, 'Yes, he is mine — I took him, a homeless little ragamuffin, but I saw what was in him, and I said his name would be heard some day — behold him, observe him — was I right?' "

The king spoke - - in a thoughtful, measured voice —

"Thou didst save me injury and shame, perchance my life, and so my crown. Such service demandeth rich reward. Name thy desire, and so it be within the compass of my royal power, it is thine."

"THOU HAST BEEN SHAMEFULLY ABUSED."

This fantastic suggestion startled Hendon out of his revery. He was about to thank the king and put the matter aside with saying he had only done his duty and desired no reward, but a wiser thought came into his head, and he asked leave to be silent a few moments and consider the gracious offer — an idea which the king gravely approved, remarking that it was best to be not too hasty with a thing of such great import.

Miles reflected during some moments, then said to himself, "Yes, that is the thing to do — by any other means it were impossible to get at it—and certes, this hour's experience has taught me 'twould be most wearing and inconvenient to continue it as it is. Yes, I will propose it; 'twas a happy accident that I did not throw the chance away." Then he dropped upon one knee and said—

"My poor service went not beyond the limit of a subject's simple duty, and therefore hath no merit; but since your majesty is pleased

"HE DROPPED ON ONE KNEE."

to hold it worthy some reward, I take heart of grace to make petition to this effect. Near four hundred years ago, as your grace knoweth, there being ill blood betwixt John, King of England, and the King of France, it was decreed that two champions should fight together in the lists, and so settle the dispute by what is called the arbitrament of

God. These two kings, and the Spanish king, being assembled to witness and judge the conflict, the French champion appeared; but so redoubtable was he, that our English knights refused to measure weapons with him. So the matter, which was a weighty one, was like to go against the English monarch by default. Now in the Tower lay the lord de Courcy, the mightiest arm in England, stripped of his honors and possessions, and wasting with long captivity. Appeal was made to him; he gave assent, and came forth arrayed for battle; but no sooner did the Frenchman glimpse his huge frame and hear his famous name but he fled away, and the French king's cause was lost. King John restored de Courcy's titles and possessions, and said, 'Name thy wish and thou shalt have it, though it cost me half my kingdom;' whereat de Courcy, kneeling, as I do now, made answer, 'This, then, I ask, my liege; that I and my successors may have and hold the privilege of remaining covered in the presence of the kings of England, henceforth while the throne shall last.' The boon was granted, as your majesty knoweth; and there hath been no time, these four hundred years, that that line has failed of an heir; and so, even unto this day, the head of that ancient house still weareth his hat or helm before the king's majesty, without let or hindrance, and this none other may do.[1] Invoking this precedent in aid of my prayer, I beseech the king to grant to me but this one grace and privilege — to my more than sufficient reward — and none other, to wit: that I and my heirs, forever, may *sit* in the presence of the majesty of England!"

"Rise, Sir Miles Hendon, Knight," said the king, gravely — giving the accolade with Hendon's sword — "rise, and seat thyself. Thy petition is granted. Whilst England remains, and the crown continues, the privilege shall not lapse."

His majesty walked apart, musing, and Hendon dropped into a chair at table, observing to himself, "'Twas a brave thought, and hath

[1] The lords of Kingsale, descendants of de Courcy, still enjoy this curious privilege.

wrought me a mighty deliverance; my legs are grievously wearied. An' I had not thought of that, I must have had to stand for weeks, till my poor lad's wits are cured." After a little, he went on, "And so I am become a knight of the Kingdom of Dreams and Shadows! A most odd and strange position, truly, for one so matter-of-fact as I. I will not laugh—no, God forbid, for this thing which is so substanceless to me is *real* to him. And to me, also, in one way, it is not a falsity, for it reflects with truth the sweet and generous spirit that is in him." After a pause: "Ah, what if he should call me by my fine title before folk!—there'd be a merry contrast betwixt my glory and my raiment! But no matter: let him call me what he will, so it please him; I shall be content."

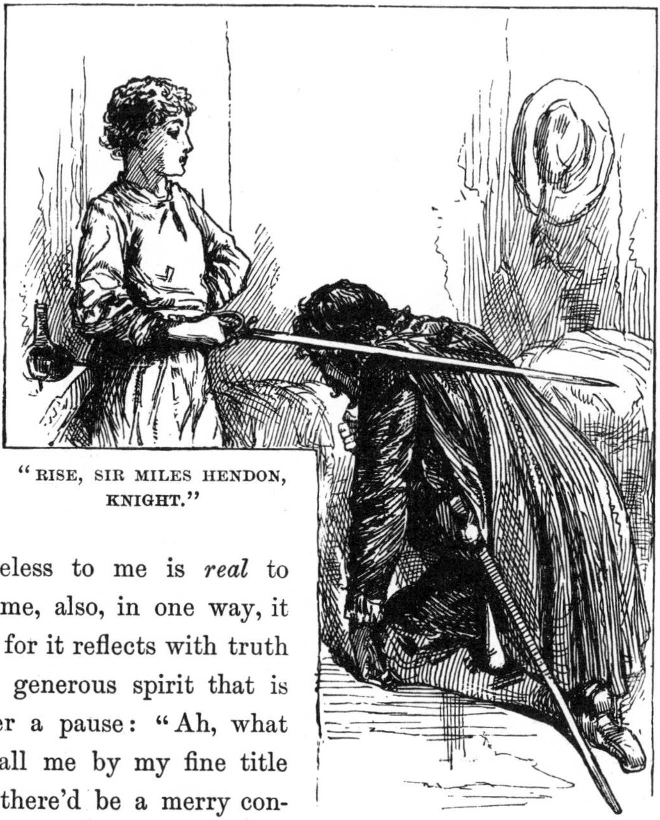

"RISE, SIR MILES HENDON, KNIGHT."

THE DISAPPEARANCE OF THE PRINCE

CHAPTER XIII.

THE DISAPPEARANCE OF THE PRINCE.

A HEAVY drowsiness presently fell upon the two comrades. The king said —

"Remove these rags" — meaning his clothing.

Hendon disapparelled the boy without dissent or remark, tucked him up in bed, then glanced about the room, saying to himself, ruefully, "He hath taken my bed again, as before — marry, what shall *I* do?" The little king observed his perplexity, and dissipated it with a word. He said, sleepily —

"Thou wilt sleep athwart the door, and guard it." In a moment more he was out of his troubles, in a deep slumber.

"Dear heart, he should have been born a king!" muttered Hendon, admiringly; "he playeth the part to a marvel."

Then he stretched himself across the door, on the floor, saying contentedly —

"I have lodged worse for seven years; 'twould be but ill gratitude to Him above to find fault with this."

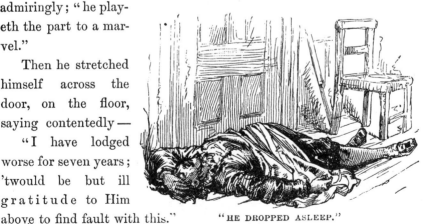

"HE DROPPED ASLEEP."

He dropped asleep as the dawn appeared. Toward noon he rose, uncovered his unconscious ward — a section at a time — and took his measure with a string. The king awoke, just as he had completed his work, complained of the cold, and asked what he was doing.

"'Tis done, now, my liege," said Hendon; "I have a bit of business outside, but will presently return; sleep thou again — thou needest it. There — let me cover thy head also — thou'lt be warm the sooner."

The king was back in dreamland before this speech was ended. Miles slipped softly out, and slipped as softly in again, in the course of thirty or forty minutes, with a complete second-hand suit of boy's clothing, of cheap material, and showing signs of wear; but tidy, and suited to the season of the year. He seated himself, and began to overhaul his purchase, mumbling to himself —

"A longer purse would have got a better sort, but when one has not the long purse one must be content with what a short one may do —

"'There was a woman in our town,
In our town did dwell' —

"He stirred, methinks — I must sing in a less thunderous key; 'tis not good to mar his sleep, with this journey before him and he so wearied out, poor chap. . . . This garment — 'tis well enough — a stitch here and another one there will set it aright. This other is better, albeit a stitch or two will not come amiss in it, likewise. . . . *These* be very good and sound, and will keep his small feet warm and dry — an odd new thing to him, belike, since he has doubtless been used to foot it bare, winters and summers the same. . . . Would thread were bread, seeing one getteth a year's sufficiency for a farthing, and such a brave big needle without cost, for mere love. Now shall I have the demon's own time to thread it!"

And so he had. He did as men have always done, and probably always will do, to the end of time — held the needle still, and tried to thrust the thread through the eye, which is the opposite of a woman's way. Time and time again the thread missed the mark, going sometimes on one side of the needle, sometimes on the other, sometimes doubling up against the shaft; but he was patient, having been through these experiences before, when he was soldiering. He succeeded at last, and took up the garment that had lain waiting, meantime, across his lap, and began his work.

"THESE BE VERY GOOD AND SOUND."

"The inn is paid — the breakfast that is to come, included — and there is wherewithal left to buy a couple of donkeys and meet our little costs for the two or three days betwixt this and the plenty that awaits us at Hendon Hall —

"'She loved her hus' —

"Body o' me! I have driven the needle under my nail! . . . It matters little — 'tis not a novelty — yet 'tis not a convenience, neither. . . . We shall be merry there, little one, never doubt it! Thy troubles will vanish, there, and likewise thy sad distemper —

"'She loved her husband dearilee,
But another man' —

"These be noble large stitches!" — holding the garment up and viewing it admiringly — "they have a grandeur and a majesty that do cause these small stingy ones of the tailor-man to look mightily paltry and plebeian —

"'She loved her husband dearilee,
But another man he loved she,' —

"Marry, 'tis done — a goodly piece of work, too, and wrought with expedition. Now will I wake him, apparel him, pour for him, feed him, and then will we hie us to the mart by the Tabard inn in Southwark and — be pleased to rise, my liege! — he answereth not — what ho, my liege! — of a truth must I profane his sacred person with a touch, sith his slumber is deaf to speech. What!"

He threw back the covers — the boy was gone!

He stared about him in speechless astonishment for a moment; noticed for the first time that his ward's ragged raiment was also missing, then he began to rage and storm, and shout for the innkeeper.— At that moment a servant entered with the breakfast.

"Explain, thou limb of Satan, or thy time is come!" roared the man of war, and made so savage a spring toward the waiter that this latter could not find his tongue, for the instant, for fright and surprise. "Where is the boy?"

In disjointed and trembling syllables the man gave the information desired.

"You were hardly gone from the place, your worship, when a youth came running and said it was your worship's will that the boy come to you straight, at the bridge-end on the Southwark side. I

"EXPLAIN, THOU LIMB OF SATAN."

brought him hither; and when he woke the lad and gave his message, the lad did grumble some little for being disturbed 'so early,' as he called it, but straightway trussed on his rags and went with the youth, only saying it had been better manners that your worship came yourself, not sent a stranger — and so" —

"And so thou'rt a fool! — a fool, and easily cozened — hang all thy breed! Yet mayhap no hurt is done. Possibly no harm is meant the boy. I will go fetch him. Make the table ready. Stay! the coverings of the bed were disposed as if one lay beneath them — happened that by accident?"

"I know not, good your worship. I saw the youth meddle with them — he that came for the boy."

"HENDON FOLLOWED AFTER HIM."

"Thousand deaths! 'twas done to deceive me — 'tis plain 'twas done to gain time. Hark ye! Was that youth alone?"

"All alone, your worship."

"Art sure?"

"Sure, your worship."

"Collect thy scattered wits — bethink thee — take time, man."

After a moment's thought, the servant said —

"When he came, none came with him; but now I remember me that as the two stepped into the throng of the Bridge, a ruffian-looking man plunged out from some near place; and just as he was joining them" —

"What *then?* — out with it!" thundered the impatient Hendon, interrupting.

"Just then the crowd lapped them up and closed them in, and I saw no more, being called by my master, who was in a rage because a joint that the scrivener had ordered was forgot, though I take all the saints to witness that to blame *me* for that miscarriage were like holding the unborn babe to judgment for sins com " —

"Out of my sight, idiot! Thy prating drives me mad! Hold! whither art flying? Canst not bide still an instant? Went they toward Southwark?"

"Even so, your worship — for, as I said before, as to that detestable joint, the babe unborn is no whit more blameless than " —

"Art here *yet!* And prating still? Vanish, lest I throttle thee!" The servitor vanished. Hendon followed after him, passed him, and plunged down the stairs two steps at a stride, muttering, "'Tis that scurvy villain that claimed he was his son. I have lost thee, my poor little mad master — it is a bitter thought — and I had come to love thee so! No! by book and bell, *not* lost! Not lost, for I will ransack the land till I find thee again. Poor child, yonder is his breakfast — and mine, but I have no hunger now — so, let the rats have it — speed, speed! that is the word!" As he wormed his swift way through the noisy multitudes upon the Bridge, he several times said to himself — clinging to the thought as if it were a particularly pleasing one — "He grumbled, but he *went* — he went, yes, because he thought Miles Hendon asked it, sweet lad — he would ne'er have done it for another, I know it well."

LE ROI
EST
MORT
VIVE LE ROI

CHAPTER XIV.

"LE ROI EST MORT—VIVE LE ROI."

TOWARD daylight of the same morning, Tom Canty stirred out of a heavy sleep and opened his eyes in the dark. He lay silent a few moments, trying to analyze his confused thoughts and impressions, and get some sort of meaning out of them, then suddenly he burst out in a rapturous but guarded voice—

"I see it all, I see it all! Now God be thanked, I am indeed awake at last! Come, joy! vanish, sorrow! Ho, Nan! Bet! kick off your straw and hie ye hither to my side, till I do pour into your unbelieving ears the wildest madcap dream that ever the spirits of night did conjure up to astonish the soul of man withal! . . . Ho, Nan, I say! Bet!" . . .

A dim form appeared at his side, and a voice said—

"Wilt deign to deliver thy commands?"

"Commands? . . . O, woe is me, I know thy voice! Speak, thou—who am I?"

"Thou? In sooth, yesternight wert thou the prince of Wales, to-day art thou my most gracious liege, Edward, King of England."

Tom buried his head among his pillows, murmuring plaintively—

"Alack, it was no dream! Go to thy rest, sweet sir—leave me to my sorrows."

Tom slept again, and after a time he had this pleasant dream. He thought it was summer and he was playing, all alone, in the fair meadow called Goodman's Fields, when a dwarf only a foot high, with long red whiskers and a humped back appeared to him suddenly and said, "Dig, by that stump." He did so, and found twelve bright new pennies — wonderful riches! Yet this was not the best of it; for the dwarf said —

"I know thee. Thou art a good lad and a deserving; thy dis-

"WILT DEIGN TO DELIVER THY COMMANDS?"

tresses shall end, for the day of thy reward is come. Dig here every seventh day, and thou shalt find always the same treasure, twelve bright new pennies. Tell none — keep the secret."

Then the dwarf vanished, and Tom flew to Offal Court with his prize, saying to himself, "Every night will I give my father a penny; he will think I begged it, it will glad his heart, and I shall no more be beaten. One penny every week the good priest that teacheth me shall have; mother, Nan and Bet the other four. We be done with hunger and rags, now, done with fears and frets and savage usage."

In his dream he reached his sordid home all out of breath, but with eyes dancing with grateful enthusiasm; cast four of his pennies into his mother's lap and cried out —

"They are for thee! — all of them, every one! — for thee and Nan and Bet — and honestly come by, not begged nor stolen!"

The happy and astonished mother strained him to her breast and exclaimed —

"It waxeth late — may it please your majesty to rise?"

Ah, that was not the answer he was expecting. The dream had snapped asunder — he was awake.

He opened his eyes — the richly clad First Lord of the Bedchamber was kneeling by his couch. The gladness of the lying dream faded away — the poor boy recognized that he was still a captive and a king. The room was filled with courtiers clothed in purple mantles — the mourning color — and with noble servants of the monarch. Tom sat up in bed and gazed out from the heavy silken curtains upon this fine company.

The weighty business of dressing began, and one courtier after another knelt and paid his court and offered to the little King his condolences upon his heavy loss, whilst the dressing proceeded. In the beginning, a shirt was taken up by the Chief Equerry in Waiting, who passed it to the First Lord of the Buckhounds, who passed it to the Second Gentleman of the Bedchamber, who passed it to the Head Ranger of Windsor Forest, who passed it to the Third Groom of the Stole, who passed it to the Chancellor Royal of the Duchy of Lancaster, who passed it to the Master of the Wardrobe, who passed it to Norroy King-at-Arms, who passed it to the Constable of the Tower, who passed it to the Chief Steward of the Household, who passed it to the Hereditary Grand Diaperer, who passed it to the Lord High Admiral of England, who passed it to the Archbishop of Canterbury, who passed it to the First Lord of the Bedchamber, who took what was left of it and put it on Tom. Poor little wondering chap, it reminded him of passing buckets at a fire.

Each garment in its turn had to go through this slow and solemn process; consequently Tom grew very weary of the ceremony; so weary that he felt an almost gushing gratefulness when he at last saw his long silken hose begin the journey down the line and knew that

"THE FIRST LORD OF THE BEDCHAMBER RECEIVED THE HOSE."

the end of the matter was drawing near. But he exulted too soon. The First Lord of the Bedchamber received the hose and was about to incase Tom's legs in them, when a sudden flush invaded his face and he hurriedly hustled the things back into the hands of the Archbishop of Canterbury with an astounded look and a whispered, "See, my lord!" — pointing to a something connected with the hose. The Archbishop paled, then flushed, and passed the hose to the Lord High Admiral, whispering, "See, my lord!" The Admiral passed the hose to the Hereditary Grand Diaperer, and had hardly breath enough in his body to ejaculate, "See, my lord!" The hose drifted backward along the line, to the Chief Steward of the Household, the Constable of the Tower, Norroy King-at-Arms, the Master of the Wardrobe, the Chancellor Royal of the Duchy of Lancaster, the Third Groom of the Stole, the Head Ranger of Windsor Forest, the Second Gentleman of the Bedchamber, the First Lord of the Buckhounds,—accompanied always with that amazed and frightened "See! see!"—till they finally reached the hands of the Chief Equerry in Waiting, who gazed a moment, with a pallid face, upon what had caused all this dismay, then hoarsely whispered, "Body of my life, a tag gone from a truss-point!—to the Tower with the Head Keeper of the King's Hose!"—after which he leaned upon the shoulder of the First Lord of the Buckhounds to regather his vanished strength whilst fresh hose, without any damaged strings to them, were brought.

But all things must have an end, and so in time Tom Canty was in a condition to get out of bed. The proper official poured water, the proper official engineered the washing, the proper official stood by with a towel, and by and by Tom got safely through the purifying stage and was ready for the services of the Hairdresser-royal. When he at length emerged from this master's hands, he was a gracious figure and as pretty as a girl, in his mantle and trunks of purple satin, and purple-plumed cap. He now moved in state toward his breakfast room, through the midst of the courtly assemblage; and as he passed, these fell back, leaving his way free, and dropped upon their knees.

After breakfast he was conducted, with regal ceremony, attended by his great officers and his guard of fifty Gentlemen Pensioners bearing gilt battle-axes, to the throne-room, where he proceeded to transact business of state. His "uncle," lord Hertford, took his stand by the throne, to assist the royal mind with wise counsel.

The body of illustrious men named by the late king as his executors, appeared, to ask Tom's approval of certain acts of theirs—rather a form, and yet not wholly a form, since there was no Protector as yet. The Archbishop of Canterbury made report of the decree of the Council of Executors concerning the obsequies of his late most illustrious majesty, and finished by reading the signatures of the Executors, to-wit: the Archbishop of Canterbury; the Lord Chancellor of

"A SECRETARY OF STATE PRESENTED AN ORDER."

England; William Lord St. John; John Lord Russell; Edward Earl of Hertford; John Viscount Lisle; Cuthbert Bishop of Durham —

Tom was not listening — an earlier clause of the document was puzzling him. At this point he turned and whispered to lord Hertford —

"What day did he say the burial hath been appointed for?"

"The 16th of the coming month, my liege."

"'Tis a strange folly. Will he keep?"

Poor chap, he was still new to the customs of royalty; he was used to seeing the forlorn dead of Offal Court hustled out of the way with a very different sort of expedition. However, the lord Hertford set his mind at rest with a word or two.

A secretary of state presented an order of the Council appointing the morrow at eleven for the reception of the foreign ambassadors, and desired the king's assent.

Tom turned an inquiring look toward Hertford, who whispered —

"Your majesty will signify consent. They come to testify their royal masters' sense of the heavy calamity which hath visited your grace and the realm of England."

Tom did as he was bidden. Another secretary began to read a preamble concerning the expenses of the late king's household, which had amounted to £28,000 during the preceding six months — a sum so vast that it made Tom Canty gasp; he gasped again when the fact appeared that £20,000 of this money were still owing and unpaid;[1] and once more when it appeared that the king's coffers were about empty, and his twelve hundred servants much embarrassed for lack of the wages due them. Tom spoke out, with lively apprehension —

"We be going to the dogs, 'tis plain. 'Tis meet and necessary that we take a smaller house and set the servants at large, sith they be of no value but to make delay, and trouble one with offices that harass the spirit and shame the soul, they misbecoming any but a doll, that hath nor brains nor hands to help itself withal. I remember me of a small house that standeth over against the fish-market, by Billingsgate" —

A sharp pressure upon Tom's arm stopped his foolish tongue and sent a blush to his face; but no countenance there betrayed any sign that this strange speech had been remarked or given concern.

[1] Hume.

A secretary made report that forasmuch as the late king had provided in his will for conferring the ducal degree upon the earl of Hertford and raising his brother, Sir Thomas Seymour, to the peerage, and likewise Hertford's son to an earldom, together with similar aggrandizements to other great servants of the crown, the Council had resolved to hold a sitting on the 16th of February for the delivering and confirming of these honors; and that meantime, the late king not having granted, in writing, estates suitable to the support of these dignities, the Council, knowing his private wishes in that regard, had thought proper to grant to Seymour "£500 lands," and to Hertford's son "800 pound lands, and 300 pound of the next bishop's lands which should fall vacant," — his present majesty being willing.[1]

Tom was about to blurt out something about the propriety of paying the late King's debts first, before squandering all this money; but a timely touch upon his arm, from the thoughtful Hertford, saved him this indiscretion; wherefore he gave the royal assent, without spoken comment, but with much inward discomfort. While he sat reflecting, a moment, over the ease with which he was doing strange and glittering miracles, a happy thought shot into his mind: why not make his mother Duchess of Offal Court and give her an estate? But a sorrowful thought swept it instantly away: he was only a king in name, these grave veterans and great nobles were his masters; to them his mother was only the creature of a diseased mind; they would simply listen to his project with unbelieving ears, then send for the doctor.

The dull work went tediously on. Petitions were read, and proclamations, patents, and all manner of wordy, repetitious and wearisome papers relating to the public business; and at last Tom sighed pathetically and murmured to himself, "In what have I offended, that the good God should take me away from the fields and the free air and the sunshine, to shut me up here and make me a king and afflict me so?"

[1] Hume.

Then his poor muddled head nodded a while, and presently drooped to his shoulder; and the business of the empire came to a stand-still for want of that august factor, the ratifying power. Silence ensued, around the slumbering child, and the sages of the realm ceased from their deliberations.

During the forenoon, Tom had an enjoyable hour, by permission of his keepers, Hertford and St. John, with the lady Elizabeth and the little lady Jane Grey; though the spirits of the princesses were rather subdued by the mighty stroke that had fallen upon the royal house; and at the end of the visit his "elder sister" — afterwards the "Bloody Mary" of history — chilled him with a solemn interview which had but one merit in his eyes, its brevity. He had a few moments to himself, and then a slim lad of about twelve years of age was admitted to his presence, whose clothing, except his snowy ruff and the laces about his wrists, was of black, — doublet, hose and all. He bore no badge of mourning but a knot of purple ribbon on his shoulder. He advanced hesitatingly, with head bowed and bare, and dropped upon one knee in front of Tom. Tom sat still and contemplated him soberly a moment. Then he said —

"Rise, lad. Who art thou? What wouldst have?"

The boy rose, and stood at graceful ease, but with an aspect of concern in his face. He said —

"Of a surety thou must remember me, my lord. I am thy whipping-boy."

"My *whipping*-boy?"

"The same, your grace. I am Humphrey — Humphrey Marlow."

Tom perceived that here was some one whom his keepers ought to have posted him about. The situation was delicate. What should he do? — pretend he knew this lad, and then betray by his every utterance, that he had never heard of him before? No, that would not do. An idea came to his relief: accidents like this might be likely to happen with some frequency, now that business urgencies would often call Hertford and St. John from his side, they being members of the council of executors; therefore perhaps it would be well to strike out a plan himself to meet the requirements of such emergencies. Yes, that would be a wise course — he would practise on this boy, and

"THE BOY ROSE, AND STOOD AT GRACEFUL EASE."

see what sort of success he might achieve. So he stroked his brow, perplexedly, a moment or two, and presently said —

"Now I seem to remember thee somewhat — but my wit is clogged and dim with suffering" —

"Alack, my poor master!" ejaculated the whipping-boy, with feeling; adding, to himself, "In truth 'tis as they said — his mind is gone — alas, poor soul! But misfortune catch me, how am I forgetting!

they said one must not seem to observe that aught is wrong with him."

"'Tis strange how my memory doth wanton with me these days," said Tom. "But mind it not — I mend apace — a little clew doth often serve to bring me back again the things and names which had escaped me. [And not they, only, forsooth, but e'en such as I ne'er heard before — as this lad shall see.] Give thy business speech."

"'Tis matter of small weight, my liege, yet will I touch upon it an' it please your grace. Two days gone by, when your majesty faulted thrice in your Greek — in the morning lessons, — dost remember it?"

"Y-e-s — methinks I do. [It is not much of a lie — an' I had meddled with the Greek at all, I had not faulted simply thrice, but forty times.] Yes, I do recall it, now — go on."

— "The master, being wroth with what he termed such slovenly and doltish work, did promise that he would soundly whip me for it — and" —

"Whip *thee!*" said Tom, astonished out of his presence of mind. "Why should he whip *thee* for faults of mine?"

"Ah, your grace forgetteth again. He always scourgeth me, when thou dost fail in thy lessons."

"True, true — I had forgot. Thou teachest me in private — then if I fail, he argueth that thy office was lamely done, and" —

"O, my liege, what words are these? I, the humblest of thy servants, presume to teach *thee?*"

"Then where is thy blame? What riddle is this? Am I in truth gone mad, or is it thou? Explain — speak out."

"But good your majesty, there's nought that needeth simplifying. — None may visit the sacred person of the prince of Wales with blows; wherefore when he faulteth, 'tis I that take them; and meet it is and right, for that it is mine office and my livelihood."[1]

[1] See Note 8, at end of volume.

Tom stared at the tranquil boy, observing to himself, "Lo, it is a wonderful thing, — a most strange and curious trade; I marvel they have not hired a boy to take my combings and my dressings for me — would heaven they would! — an' they will do this thing, I will take my lashings in mine own person, giving God thanks for the change." Then he said aloud —

"And hast thou been beaten, poor friend, according to the promise?"

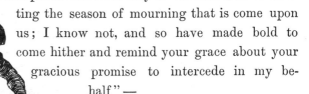

"No, good your majesty, my punishment was appointed for this day, and peradventure it may be annulled, as unbefitting the season of mourning that is come upon us; I know not, and so have made bold to come hither and remind your grace about your gracious promise to intercede in my behalf" —

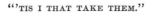

"'TIS I THAT TAKE THEM."

"With the master? To save thee thy whipping?"

"Ah, thou dost remember!"

"My memory mendeth, thou seest. Set thy mind at ease — thy back shall go unscathed — I will see to it."

"O, thanks, my good lord!" cried the boy, dropping upon his knee again. "Mayhap I have ventured far enow; and yet" . . .

Seeing Master Humphrey hesitate, Tom encouraged him to go on, saying he was "in the granting mood."

"Then will I speak it out, for it lieth near my heart. Sith thou art no more prince of Wales but King, thou canst order matters as thou wilt, with none to say thee nay; wherefore it is not in reason that thou wilt longer vex thyself with dreary studies, but wilt burn thy books and turn thy mind to things less irksome. Then am I ruined, and mine orphan sisters with me!"

"Ruined? Prithee how?"

"My back is my bread, O my gracious liege! if it go idle, I starve. An' thou cease from study, mine office is gone, thou'lt need no whipping-boy. Do not turn me away!"

Tom was touched with this pathetic distress. He said, with a right royal burst of generosity —

"Discomfort thyself no further, lad. Thine office shall be permanent in thee and thy line, forever." Then he struck the boy a light blow on the shoulder with the flat of his sword, exclaiming, "Rise, Humphrey Marlow, Hereditary Grand Whipping-Boy to the royal house of England! Banish sorrow — I will betake me to my books again, and study so ill that they must in justice treble thy wage, so mightily shall the business of thine office be augmented."

The grateful Humphrey responded fervidly —

"Thanks, O most noble master, this princely lavishness doth far surpass my most distempered dreams of fortune. Now shall I be happy all my days, and all the house of Marlow after me."

Tom had wit enough to perceive that here was a lad who could be useful to him. He encouraged Humphrey to talk, and he was nothing loath. He was delighted to believe that he was helping in Tom's "cure;" for always, as soon as he had finished calling back to Tom's diseased mind the various particulars of his experiences and adventures in the royal school-room and elsewhere about the palace, he noticed that Tom was then able to "recall" the circumstances quite clearly. At the end of an hour Tom found himself well freighted with very valuable information concerning personages and matters pertaining to the Court; so he resolved to draw instruction from this source daily; and to this end he would give order to admit Humphrey to the royal closet whenever he might come, provided the majesty of England was not engaged with other people. Humphrey had hardly been dismissed when my lord Hertford arrived with more trouble for Tom.

He said that the lords of the Council, fearing that some overwrought report of the king's damaged health might have leaked out and got abroad, they deemed it wise and best that his majesty should begin to dine in public after a day or two — his wholesome complexion and vigorous step, assisted by a carefully guarded repose of manner and ease and grace of demeanor, would more surely quiet the general pulse — in case any evil rumors *had* gone about — than any other scheme that could be devised.

Then the earl proceeded, very delicately, to instruct Tom as to the observances proper to the stately occasion, under the rather thin disguise of "reminding" him concerning things already known to him; but to his vast gratification it turned out that Tom needed very little help in this line — he had been making use of Humphrey in that direction, for Humphrey had mentioned that within a few days he was to begin to dine in public; having gathered it from the swift-winged gossip of the Court. Tom kept these facts to himself, however.

Seeing the royal memory so improved, the earl ventured to apply a few tests to it, in an apparently casual way, to find out how far its amendment had progressed. The results were happy, here and there, in spots — spots where Humphrey's tracks remained — and on the whole my lord was greatly pleased and encouraged. So encouraged was he, indeed, that he spoke up and said in a quite hopeful voice —

"Now am I persuaded that if your majesty will but tax your memory yet a little further, it will resolve the puzzle of the Great Seal — a loss which was of moment yesterday, although of none to-day, since its term of service ended with our late lord's life. May it please your grace to make the trial?"

Tom was at sea — a Great Seal was a something which he was

totally unacquainted with. After a moment's hesitation he looked up innocently and asked —

"What was it like, my lord?"

The earl started, almost imperceptibly, muttering to himself,

"IF YOUR MAJESTY WILL BUT TAX YOUR MEMORY."

"Alack, his wits are flown again! — it was ill wisdom to lead him on to strain them" — then he deftly turned the talk to other matters, with the purpose of sweeping the unlucky Seal out of Tom's thoughts — a purpose which easily succeeded.

TOM·AS
·KING·

CHAPTER XV.

TOM AS KING.

THE next day the foreign ambassadors came, with their gorgeous trains; and Tom, throned in awful state, received them. The splendors of the scene delighted his eye and fired his imagination, at first, but the audience was long and dreary, and so were most of the addresses — wherefore, what began as a pleasure, grew into weariness and homesickness by and by. Tom said the words which Hertford put into his mouth from time to time, and tried hard to acquit himself satisfactorily, but he was too new to such things, and too ill at ease to accomplish more than a tolerable success. He looked sufficiently like a king, but he was ill able to feel like one. He was cordially glad when the ceremony was ended.

The larger part of his day was "wasted" — as he termed it, in his own mind — in labors pertaining to his royal office. Even the two hours devoted to certain princely pastimes and recreations were rather a burden to him, than otherwise, they were so fettered by restrictions and ceremonious observances. However he had a private hour with his whipping-boy which he counted clear gain, since he got both entertainment and needful information out of it.

The third day of Tom Canty's Kingship came and went much as the others had done, but there was a lifting of his cloud in one way — he felt less uncomfortable than at first; he was getting a little used to his circumstances and surroundings; his chains still galled, but not all the time; he found that the presence and homage of the great afflicted and embarrassed him less and less sharply with every hour that drifted over his head.

But for one single dread, he could have seen the fourth day approach without serious distress — the dining in public; it was to begin that day. There were greater matters in the programme — for on that day he would have to preside at a Council which would take his views and commands concerning the policy to be pursued toward various foreign nations scattered far and near over the great globe; on that day, too, Hertford would be formally chosen to the grand office of Lord Protector; other things of note were appointed for that fourth day, also; but to Tom they were all insignificant compared with the ordeal of dining all by himself with a multitude of curious eyes fastened upon him and a multitude of mouths whispering comments upon his performance, — and upon his mistakes, if he should be so unlucky as to make any.

Still, nothing could stop that fourth day, and so it came. It found poor Tom low-spirited and absent-minded, and this mood continued; he could not shake it off. The ordinary duties of the morning dragged upon his hands, and wearied him. Once more he felt the sense of captivity heavy upon him.

Late in the forenoon he was in a large audience chamber, conversing with the earl of Hertford and dully awaiting the striking of the hour appointed for a visit of ceremony from a considerable number of great officials and courtiers.

After a little while, Tom, who had wandered to a window and become interested in the life and movement of the great highway beyond the palace gates — and not idly interested, but longing with all his heart to take part in person in its stir and freedom — saw the van of a hooting and shouting mob of disorderly men, women and children of the lowest and poorest degree approaching from up the road.

"I would I knew what 'tis about!" he exclaimed, with all a boy's curiosity in such happenings.

"Thou art the king!" solemnly responded the earl, with a reverence. "Have I your grace's leave to act?"

"O blithely, yes! O gladly yes!" exclaimed Tom, excitedly, adding to himself with a lively sense of satisfaction, "In truth, being a king is not all dreariness — it hath its compensations and conveniences."

"TOM HAD WANDERED TO A WINDOW."

The earl called a page, and sent him to the captain of the guard with the order —

"Let the mob be halted, and inquiry made concerning the occasion of its movement. By the king's command!"

A few seconds later a long rank of the royal guards, cased in flashing steel, filed out at the gates and formed across the highway in front of the multitude. A messenger returned, to report that the crowd were following a man, a woman, and a young girl to execution for crimes committed against the peace and dignity of the realm.

Death — and a violent death — for these poor unfortunates! The thought wrung Tom's heart-strings. The spirit of compassion took control of him, to the exclusion of all other considerations; he never thought of the offended laws, or of the grief or loss which these three criminals had inflicted upon their victims, he could think of nothing but the scaffold and the grisly fate hanging over the heads of the condemned. His concern made him even forget, for the moment, that he was but the false shadow of a king, not the substance; and before he knew it he had blurted out the command—

"Bring them here!"

Then he blushed scarlet, and a sort of apology sprung to his lips; but observing that his order had wrought no sort of surprise in the earl or the waiting page, he suppressed the words he was about to utter. The page, in the most matter-of-course way, made a profound obeisance and retired backwards out of the room to deliver the command. Tom experienced a glow of pride and a renewed sense of the compensating advantages of the kingly office. He said to himself, "Truly it is like what I was used to feel when I read the old priest's tales, and did imagine mine own self a prince, giving law and command to all, saying 'Do this, do that,' whilst none durst offer let or hindrance to my will."

Now the doors swung open; one high-sounding title after another was announced, the personages owning them followed, and the place was quickly half filled with noble folk and finery. But Tom was hardly conscious of the presence of these people, so wrought up was he and so intensely absorbed in that other and more interesting matter. He seated himself, absently, in his chair of state, and turned his eyes upon the door with manifestations of impatient expectancy; seeing which, the company forebore to trouble him, and fell to chatting a mixture of public business and court gossip one with another.

In a little while the measured tread of military men was heard approaching, and the culprits entered the presence in charge of an under-sheriff and escorted by a detail of the king's guard. The civil officer knelt before Tom, then stood aside; the three doomed persons knelt, also, and remained so; the guard took position behind Tom's chair. Tom scanned the prisoners curiously. Something about the dress or appearance of the man had stirred a vague memory in him. "Methinks I have seen this man ere now . . . but the when or the where fail me" — such was Tom's thought. Just then the man glanced quickly up, and quickly dropped his face again, not being able to endure the awful port of sovereignty; but the one full glimpse of the face, which Tom got, was sufficient. He said to himself: "Now is the matter clear; this is the stranger that plucked Giles Witt out of the Thames, and saved his life, that windy, bitter, first day of the New Year — a brave good deed — pity he hath been doing baser

"TOM SCANNED THE PRISONERS."

ones and got himself in this sad case. . . . I have not forgot the day, neither the hour; by reason that an hour after, upon the stroke of eleven, I did get a hiding by the hand of Gammer Canty which was of so goodly and admired severity that all that went before or followed after it were but fondlings and caresses by comparison."

Tom now ordered that the woman and the girl be removed from the presence for a little time; then addressed himself to the under-sheriff, saying —

"Good sir, what is this man's offence?"

The officer knelt, and answered —

"So please your majesty, he hath taken the life of a subject by poison."

Tom's compassion for the prisoner, and admiration of him as the daring rescuer of a drowning boy, experienced a most damaging shock.

"The thing was proven upon him?" he asked.

"Most clearly, sire."

Tom sighed, and said —

"Take him away — he hath earned his death. 'Tis a pity, for he was a brave heart — na — na, I mean he hath the *look* of it!"

The prisoner clasped his hands together with sudden energy, and wrung them despairingly, at the same time appealing imploringly to the "king" in broken and terrified phrases —

"O my lord the king, an' thou canst pity the lost, have pity upon me! I am innocent — neither hath that wherewith I am charged been more than but lamely proved — yet I speak not of that; the judgment is gone forth against me and may not suffer alteration; yet in mine extremity I beg a boon, for my doom is more than I can bear. A grace, a grace, my lord the king! in thy royal compassion grant my prayer — give commandment that I be hanged!"

Tom was amazed. This was not the outcome he had looked for.

"Odds my life, a strange *boon!* Was it not the fate intended thee?"

"O good my liege, not so! It is ordered that I be *boiled alive!*"

The hideous surprise of these words almost made Tom spring from his chair. As soon as he could recover his wits he cried out —

"Have thy wish, poor soul! an' thou had poisoned a hundred men thou shouldst not suffer so miserable a death."

The prisoner bowed his face to the ground and burst into passionate expressions of gratitude — ending with —

"If ever thou shouldst know misfortune — which God forefend! — may thy goodness to me this day be remembered and requited!"

Tom turned to the earl of Hertford, and said —

"My lord, is it believable that there was warrant for this man's ferocious doom?"

"It is the law, your grace — for poisoners. In Germany coiners be boiled to death in *oil* — not cast in of a sudden, but by a rope let down into the oil by degrees, and slowly; first the feet, then the legs, then" —

"O prithee no more, my lord, I cannot bear it!" cried Tom, covering his eyes with his hands to shut out the picture. "I beseech your good lordship that order be taken to change this law — O, let no more poor creatures be visited with its tortures."

The earl's face showed profound gratification, for he was a man of merciful and generous impulses — a thing not very common with his class in that fierce age. He said —

"These your grace's noble words have sealed its doom. History will remember it to the honor of your royal house."

The under-sheriff was about to remove his prisoner; Tom gave him a sign to wait; then he said —

"Good sir, I would look into this matter further. The man has said his deed was but lamely proved. Tell me what thou knowest."

"If the king's grace please, it did appear upon the trial, that this man entered into a house in the hamlet of Islington where one lay sick — three witnesses say it was at ten of the clock in the morning and two say it was some minutes later — the sick man being alone at the time, and sleeping — and presently the man came forth again, and went his way. The sick man died within the hour, being torn with spasms and retchings."

"Did any see the poison given? Was poison found?"

"Marry, no, my liege."

"Then how doth one know there was poison given at all?"

"Please your majesty, the doctors testified that none die with such symptoms but by poison."

Weighty evidence, this — in that simple age. Tom recognized its formidable nature, and said —

"The doctor knoweth his trade — belike they were right. The matter hath an ill look for this poor man."

"Yet was not this all, your majesty; there is more and worse. Many testified that a witch, since gone from the village, none know whither, did foretell, and speak it privately in their ears, that the sick man *would die by poison* — and more, that a stranger would give it — a stranger with brown hair and clothed in a worn and common garb; and surely this prisoner doth answer woundily to the bill. Please your majesty to give the circumstance that solemn weight which is its due, seeing it was *foretold*."

This was an argument of tremendous force, in that superstitious day. Tom felt that the thing was settled; if evidence was worth any thing, this poor fellow's guilt was proved. Still he offered the prisoner a chance, saying —

"If thou canst say aught in thy behalf, speak."

"Nought that will avail, my king. I am innocent, yet cannot I make it appear. I have no friends, else might I show that I was not in Islington that day; so also might I show that at that hour they name I was above a league away, seeing I was at Wapping Old Stairs; yea more, my King, for I could show, that whilst they say I was *taking* life, I was *saving* it. A drowning boy" —

"Peace! Sheriff, name the day the deed was done!"

"At ten in the morning, or some minutes later, the first day of the new year, most illustrious" —

"Let the prisoner go free — it is the king's will!"

Another blush followed this unregal outburst, and he covered his indecorum as well as he could by adding —

"It enrageth me that a man should be hanged upon such idle, hare-brained evidence!"

A low buzz of admiration swept through the assemblage. It was not admiration of the decree that had been delivered by Tom, for the propriety or expediency of pardoning a convicted poisoner

"LET THE PRISONER GO FREE!"

was a thing which few there would have felt justified in either admitting or admiring — no, the admiration was for the intelligence and spirit which

Tom had displayed. Some of the low-voiced remarks were to this effect —

"This is no mad king — he hath his wits sound."

"How sanely he put his questions — how like his former natural self was this abrupt, imperious disposal of the matter!"

"God be thanked his infirmity is spent! This is no weakling, but a king. He hath borne himself like to his own father."

The air being filled with applause, Tom's ear necessarily caught a little of it. The effect which this had upon him was to put him greatly at his ease, and also to charge his system with very gratifying sensations.

However, his juvenile curiosity soon rose superior to these pleasant thoughts and feelings; he was eager to know what sort of deadly mischief the woman and the little girl could have been about; so, by his command the two terrified and sobbing creatures were brought before him.

"What is it that these have done?" he inquired of the sheriff.

"WHAT IS IT THAT THESE HAVE DONE?"

"Please your majesty, a black crime is charged upon them, and clearly proven; wherefore the judges have decreed, according to the law, that they be hanged. They sold themselves to the devil — such is their crime."

Tom shuddered. He had been taught to abhor people who did this wicked thing. Still, he was not going to deny himself the pleasure of feeding his curiosity, for all that; so he asked —

"Where was this done? — and when?"

"On a midnight, in December — in a ruined church, your majesty."

Tom shuddered again.

"Who was there present?"

"Only these two, your grace — and *that other*."

"Have these confessed?"

"Nay, not so, sire — they do deny it."

"Then prithee, how was it known?"

"Certain witnesses did see them wending thither, good your majesty; this bred the suspicion, and dire effects have since confirmed and justified it. In particular, it is in evidence that through the wicked power so obtained, they did invoke and bring about a storm that wasted all the region round about. Above forty witnesses have proved the storm; and sooth one might have had a thousand, for all had reason to remember it, sith all had suffered by it."

"Certes this is a serious matter." Tom turned this dark piece of scoundrelism over in his mind a while, then asked —

"Suffered the woman, also, by the storm?"

Several old heads among the assemblage nodded their recognition of the wisdom of this question. The sheriff, however, saw nothing consequential in the inquiry; he answered, with simple directness —

"Indeed, did she, your majesty, and most righteously, as all aver. Her habitation was swept away, and herself and child left shelterless."

"Methinks the power to do herself so ill a turn was dearly bought. She had been cheated, had she paid but a farthing for it; that she paid her soul, and her child's, argueth that she is mad; if she is mad she knoweth not what she doth, therefore sinneth not."

The elderly heads nodded recognition of Tom's wisdom once more, and one individual murmured, "An' the king be mad himself, according to report, then is it a madness of a sort that would improve the sanity of some I wot of, if by the gentle providence of God they could but catch it."

"What age hath the child?" asked Tom.

"Nine years, please your majesty."

"By the law of England may a child enter into covenant and sell itself, my lord?" asked Tom, turning to a learned judge.

"The law doth not permit a child to make or meddle in any weighty matter, good my liege, holding that its callow wit unfitteth it to cope with the riper wit and evil schemings of them that are its elders. The *devil* may buy a child, if he so choose, and the child agree thereto, but not an Englishman — in this latter case the contract would be null and void."

"It seemeth a rude unchristian thing, and ill contrived, that

"SEVERAL OLD HEADS NODDED THEIR RECOGNITION."

English law denieth privileges to Englishmen, to waste them on the devil!" cried Tom, with honest heat.

This novel view of the matter excited many smiles, and was stored away in many heads to be repeated about the court as evidence of Tom's originality as well as progress toward mental health.

The elder culprit had ceased from sobbing, and was hanging upon Tom's words with an excited interest and a growing hope. Tom noticed this, and it strongly inclined his sympathies toward her in her perilous and unfriended situation. Presently he asked —

"How wrought they, to bring the storm?"

"*By pulling off their stockings*, sire."

This astonished Tom, and also fired his curiosity to fever heat. He said, eagerly —

"It is wonderful! Hath it always this dread effect?

"Always, my liege — at least if the woman desire it, and utter the needful words, either in her mind or with her tongue."

Tom turned to the woman, and said with impetuous zeal —

"Exert thy power — I would see a storm!"

There was a sudden paling of cheeks in the superstitious assemblage, and a general, though unexpressed, desire to get out of the place — all of which was lost upon Tom, who was dead to every thing but the proposed cataclysm. Seeing a puzzled and astonished look in the woman's face, he added, excitedly —

"Never fear — thou shalt be blameless. More — thou shalt go free — none shall touch thee. Exert thy power."

"O, my lord the king, I have it not — I have been falsely accused."

"Thy fears stay thee. Be of good heart, thou shalt suffer no harm. Make a storm — it mattereth not how small a one — I require nought great or harmful, but indeed prefer the opposite — do this and thy life is spared — thou shalt go out free, with thy child, bearing the king's pardon, and safe from hurt or malice from any in the realm."

The woman prostrated herself, and protested, with tears, that she had no power to do the miracle, else she would gladly win her child's life, alone, and be content to lose her own, if by obedience to the king's command so precious a grace might be acquired.

Tom urged — the woman still adhered to her declarations. Finally he said —

"I think the woman hath said true. An' *my* mother were in her place and gifted with the devil's functions, she had not stayed a moment to call her storms and lay the whole land in ruins, if the saving of my forfeit life were the price she got! It is argument that other mothers are made in like mould. Thou art free, goodwife — thou and thy child — for I do think thee innocent. *Now* thou'st nought to fear, being pardoned — pull off thy stockings! — an' thou canst make me a storm, thou shalt be rich!"

The redeemed creature was loud in her gratitude, and proceeded to obey, whilst Tom looked on with eager expectancy, a little marred by apprehension; the courtiers at the same time manifesting decided discomfort and uneasiness. The woman stripped her own feet and her little girl's also, and plainly did her best to reward the king's generosity with an earthquake, but it was all a failure and a disappointment. Tom sighed, and said —

"There, good soul, trouble thyself no further, thy power is departed out of thee. Go thy way in peace; and if it return to thee at any time, forget me not, but fetch me a storm."[1]

[1] See Notes to Chapter xv at the end of the volume.

THE
STATE
DINNER

CHAPTER XVI.

THE STATE DINNER.

THE dinner hour drew near — yet strangely enough, the thought brought but slight discomfort to Tom, and hardly any terror. The morning's experiences had wonderfully built up his confidence; the poor little ash-cat was already more wonted to his strange garret, after four days' habit, than a mature person could have become in a full month. A child's facility in accommodating itself to circumstances was never more strikingly illustrated.

Let us privileged ones hurry to the great banqueting room and have a glance at matters there whilst Tom is being made ready for the imposing occasion. It is a spacious apartment, with gilded pillars and pilasters, and pictured walls and ceilings. At the door stand tall guards, as rigid as statues, dressed in rich and picturesque costumes, and bearing halberds. In a high gallery which runs all around the place is a band of musicians and a packed company of citizens of both sexes, in brilliant attire. In the centre of the room, upon a raised platform, is Tom's table. Now let the ancient chronicler speak:

"A gentleman enters the room bearing a rod, and along with him another bearing a table-cloth, which, after they have both kneeled three times with the utmost veneration, he spreads upon the table, and after kneeling again they both retire; then come two others, one with the rod again, the other with a salt-cellar, a plate, and bread; when they have kneeled as the others had done, and placed what was brought upon the table, they too retire with the same ceremonies performed by the first; at last come two nobles, richly clothed, one bearing a tasting-knife, who, after prostrating themselves three times in the most graceful manner, approach and rub the table with bread and salt, with as much awe as if the king had been present." [1]

So end the solemn preliminaries. Now, far down the echoing corridors we hear a bugle-blast, and the indistinct cry, "Place for the king! way for the king's most excellent majesty!" These sounds are momently repeated — they grow nearer and nearer — and presently, almost in our faces, the martial note peals and the cry rings out, "Way for the king!" At this instant the shining pageant appears, and files in at the door, with a measured march. Let the chronicler speak again:

"A GENTLEMAN BEARING A ROD."

"First come Gentlemen, Barons, Earls, Knights of the Garter, all richly dressed and bareheaded; next comes the Chancellor, between two, one of which carries the royal sceptre, the other the Sword of

[1] Leigh Hunt's "The Town," p. 408, quotation from an early tourist.

State in a red scabbard, studded with golden fleurs-de-lis, the point upwards; next comes the King himself — whom, upon his appearing, twelve trumpets and many drums salute with a great burst of welcome, whilst all in the galleries rise in their places, crying "God save the King!" After him come nobles attached to his person, and on his right and left march his guard of honor, his fifty Gentlemen Pensioners, with gilt battle-axes."

This was all fine and pleasant. Tom's pulse beat high and a glad light was in his eye. He bore himself right gracefully, and all the more so because he was not thinking of how he was doing it, his mind being charmed and occupied with the blithe sights and sounds about him — and besides, nobody can be very ungraceful in nicely-fitting beautiful clothes after he has grown a little used to them — especially if he is for the moment unconscious of them.

"THE CHANCELLOR BETWEEN TWO."

Tom remembered his instructions, and acknowledged his greeting with a slight inclination of his plumed head, and a courteous "I thank ye, my good people."

He seated himself at table, without removing his cap; and did it without the least embarrassment; for to eat with one's cap on was the one solitary royal custom upon which the kings and the Cantys met upon common ground, neither party having any advantage over the other in the matter of old familiarity with it. The pageant broke up and grouped itself picturesquely, and remained bareheaded.

"I THANK YE, MY GOOD PEOPLE."

Now, to the sound of gay music, the Yeomen of the Guard entered, — "the tallest and mightiest men in England, they being carefully selected in this regard" — but we will let the chronicler tell about it:

"The Yeoman of the Guard entered, bare-headed, clothed in scarlet, with golden roses upon their backs; and these went and came, bringing in each turn a course of dishes, served in plate. These dishes were received by a gentleman in the same order they were brought, and placed upon the table, while the taster gave to each guard a mouthful to eat of the particular dish he had brought, for fear of any poison."

Tom made a good dinner, notwithstanding he was conscious that hundreds of eyes followed each morsel to his mouth and watched him eat it with an interest which could not have been more intense if it had been a deadly explosive and was expected to blow him up and scatter him all about the place. He was careful not to hurry, and equally careful not to do any thing whatever for himself, but wait till the proper official knelt down and did it for him. He got through without a mistake — flawless and precious triumph.

"HE MARCHED AWAY IN THE MIDST OF HIS PAGEANT."

When the meal was over at last and he marched away in the midst of his bright pageant, with the happy noises in his ears of blaring bugles, rolling drums and thundering acclamations, he felt that if he had seen the worst of dining in public, it was an ordeal which he would be glad to endure several times a day if by that means he could but buy himself free from some of the more formidable requirements of his royal office.

FOO-FOO THE FIRST

CHAPTER XVII.

FOO-FOO THE FIRST.

MILES HENDON hurried along toward the Southwark end of the bridge, keeping a sharp lookout for the persons he sought, and hoping and expecting to overtake them presently. He was disappointed in this, however. By asking questions, he was enabled to track them part of the way through Southwark; then all traces ceased, and he was perplexed as to how to proceed. Still, he continued his efforts as best he could during the rest of the day. Nightfall found him leg-weary, half famished, and his desire as far from accomplishment as ever; so he supped at the Tabard inn and went to bed, resolved to make an early start in the morning, and give the town an exhaustive search. As he lay thinking and planning, he presently began to reason thus: The boy would escape from the ruffian, his reputed father, if possible; would he go back to London and seek his former haunts? no, he would not do that, he would avoid recapture. What, then, would he do? Never having had a friend in the world, or a protector, until he met Miles Hendon, he would naturally try to find that friend again, provided the effort did not require him to go toward London and danger. He would strike for Hendon Hall, that is what he would do, for he knew Hendon was homeward bound and there he might expect to find him. Yes, the case was plain to Hendon — he must lose no more time in Southwark, but move at once through Kent, toward

Monk's Holm, searching the wood and inquiring as he went. Let us return to the vanished little king, now.

The ruffian whom the waiter at the inn on the bridge saw "about to join" the youth and the king, did not exactly join them, but fell in close behind them and followed their steps. He said nothing. His left arm was in a sling, and he wore a large green patch over his left eye; he limped slightly, and used an oaken staff as a support. The youth led the king a crooked course through Southwark, and by and by struck into the high road beyond. The king was irritated, now, and said he would stop here — it was Hendon's place to come to him, not his to go to Hendon. He would not endure such insolence; he would stop where he was. The youth said —

"Thou'lt tarry here, and thy friend lying wounded in the wood yonder? So be it, then."

The king's manner changed at once. He cried out —

"Wounded? And who hath dared to do it? But that is apart; lead on, lead on! Faster, sirrah! art shod with lead? Wounded, is he? Now though the doer of it be a duke's son, he shall rue it!"

It was some distance to the wood, but the space was speedily traversed. The youth looked about him, discovered a bough sticking in the ground, with a small bit of rag tied to it, then led the way into the forest, watching for similar boughs and finding them at intervals: they were evidently guides to the point he was aiming at. By and by an open place was reached, where were the charred remains of a farm house, and near them a barn which was falling to ruin and decay. There was no sign of life anywhere, and utter silence prevailed. The youth entered the barn, the king following eagerly upon his heels. No one there! The king shot a surprised and suspicious glance at the youth, and asked —

"Where is he?"

A mocking laugh was his answer. The king was in a rage in a moment; he seized a billet of wood and was in the act of charging upon the youth when another mocking laugh fell upon his ear. It was from the lame ruffian, who had been following at a distance. The king turned and said angrily —

"Who art thou? What is thy business here?"

"Leave thy foolery," said the man, "and quiet thyself. My disguise is none so good that thou canst pretend thou knowest not thy father through it."

"Thou art not my father. I know thee not. I am the king. If thou hast hid my servant, find him for me, or thou shalt sup sorrow for what thou hast done."

"THE RUFFIAN FOLLOWED THEIR STEPS."

John Canty replied, in a stern and measured voice —

"It is plain thou art mad, and I am loath to punish thee; but if thou provoke me, I must. Thy prating doth no harm here, where there are no ears that need to mind thy follies, yet is it well to practise thy tongue to wary speech, that it may do no hurt when our quarters change. I have done a murder, and may not tarry at home — neither shalt thou, seeing I need thy service. My name is changed, for wise reasons; it is Hobbs — John Hobbs; thine is Jack — charge thy memory accordingly. Now, then, speak.

Where is thy mother? where are thy sisters? They came not to the place appointed — knowest thou whither they went?"

The king answered, sullenly —

"Trouble me not with these riddles. My mother is dead; my sisters are in the palace."

"HE SEIZED A BILLET OF WOOD."

The youth near by burst into a derisive laugh, and the king would have assaulted him, but Canty — or Hobbs, as he now called himself — prevented him, and said —

"Peace, Hugo, vex him not; his mind is astray, and thy ways fret him. Sit thee down, Jack, and quiet thyself; thou shalt have a morsel to eat, anon."

Hobbs and Hugo fell to talking together, in low voices, and the king removed himself as far as he could from their disagreeable company. He withdrew into the twilight of the farther end of the barn, where he found the earthen floor bedded a foot deep with straw. He lay down here, drew straw over himself in lieu of blankets, and was

"HE WAS SOON ABSORBED IN THINKING."

soon absorbed in thinkings. He had many griefs, but the minor ones were swept almost into forgetfulness by the supreme one, the loss of his father. To the rest of the world the name of Henry VIII. brought a shiver, and suggested an ogre whose nostrils breathed destruction and whose hand dealt scourgings and death; but to this boy the name brought only sensations of pleasure, the figure it invoked wore a countenance that was all gentleness and affection. He called to mind a long succession of loving passages between his father and himself, and dwelt fondly upon them, his unstinted tears attesting how deep and real was the grief that possessed his heart. As the afternoon wasted away, the lad, wearied with his troubles, sunk gradually into a tranquil and healing slumber.

After a considerable time — he could not tell how long — his senses

"A GRIM AND UNSIGHTLY PICTURE."

struggled to a half-consciousness, and as he lay with closed eyes vaguely wondering where he was and what had been happening, he noted a murmurous sound, the sullen beating of rain upon the roof. A snug sense of comfort stole over him, which was rudely broken, the next moment, by a chorus of piping cackles and coarse laughter. It startled him disagreeably, and he unmuffled his head to see whence this interruption proceeded. A grim and unsightly picture met his eye. A bright fire was burning in the middle of the floor, at the other end of the barn; and around it, and lit weirdly up by the red glare, lolled and sprawled the motliest company of tattered gutter-scum and ruffians, of both sexes, he had ever read or dreamed of. There were huge, stalwart men, brown with exposure, long-haired, and clothed in fantastic rags; there were middle-sized youths, of truculent countenance, and similarly clad; there were blind mendicants, with patched or bandaged eyes; crippled ones, with wooden legs and crutches; there was a villain-looking peddler with his pack; a knife-grinder, a tinker, and a barber-surgeon, with the implements of their trades; some of the females were hardly-grown girls, some were at prime, some were old and wrinkled hags, and all were loud, brazen, foul-mouthed; and all soiled and slatternly; there were three sore-faced babies; there were a couple of starveling curs, with strings about their necks, whose office was to lead the blind.

The night was come, the gang had just finished feasting, an orgy was beginning; the can of liquor was passing from mouth to mouth. A general cry broke forth —

"A song! a song from the Bat and Dick Dot-and-go-One!"

One of the blind men got up, and made ready by casting aside the patches that sheltered his excellent eyes, and the pathetic placard which recited the cause of his calamity. Dot-and-go-One disencumbered himself of his timber leg and took his place, upon sound and healthy limbs, beside his fellow-rascal; then they roared out a rollicking ditty, and were re-enforced by the whole crew, at the end of each stanza, in a rousing chorus. By the time the last stanza was reached, the half-drunken enthusiasm had risen to such a pitch, that everybody joined in and sang it clear through from the beginning, producing a volume of villanous sound that made the rafters quake. These were the inspiring words:

"Bien Darkmans then, Bouse Mort and Ken,
The bien Coves bings awast,
On Chates to trine by Rome Coves dine
For his long lib at last.
Bing'd out bien Morts and toure, and toure,
Bing out of the Rome vile bine,
And toure the Cove that cloy'd your duds,
Upon the Chates to trine."[1]

Conversation followed; not in the thieves' dialect of the song, for that was only used in talk when unfriendly ears might be listening. In the course of it it appeared that "John Hobbs" was not altogether a new recruit, but had trained in the gang at some former time. His later history was called for, and when he said he had "accidentally" killed a man, considerable satisfaction was expressed; when he added that the man was a priest, he was roundly applauded, and had to take a drink with everybody. Old acquaintances welcomed him joyously, and new ones were proud to shake him by the hand. He was asked why he had "tarried away so many months." He answered —

"THEY ROARED OUT A ROLLICKING DITTY."

"London is better than the country, and safer these late

[1] From "The English Rogue;" London, 1665.

years, the laws be so bitter and so diligently enforced. An' I had not had that accident, I had staid there. I had resolved to stay, and never more venture country-wards — but the accident has ended that."

He inquired how many persons the gang numbered now. The "Ruffler," or chief, answered —

"Five and twenty sturdy budges, bulks, files, clapperdogeons and maunders, counting the dells and doxies and other morts.[1] Most are here, the rest are wandering eastward, along the winter lay. We follow at dawn."

"I do not see the Wen among the honest folk about me. Where may he be?"

"Poor lad, his diet is brimstone, now, and over hot for a delicate taste. He was killed in a brawl, somewhere about midsummer."

"I sorrow to hear that; the Wen was a capable man, and brave."

"That was he, truly. Black Bess, his dell, is of us yet, but absent on the eastward tramp; a fine lass, of nice ways and orderly conduct, none ever seeing her drunk above four days in the seven."

"She was ever strict — I remember it well — a goodly wench and worthy all commendation. Her mother was more free and less particular; a troublesome and ugly tempered beldame, but furnished with a wit above the common."

"We lost her through it. Her gift of palmistry and other sorts of fortune-telling begot for her at last a witch's name and fame. The law roasted her to death at a slow fire. It did touch me to a sort of tenderness to see the gallant way she met her lot — cursing and reviling all the crowd that gaped and gazed around her, whilst the flames licked upward toward her face and catched her thin locks and crackled about her old gray head — cursing them, said I? — cursing them!

[1] Canting terms for various kinds of thieves, beggars and vagabonds, and their female companions.

why an' thou shouldst live a thousand years thoud'st never hear so masterful a cursing. Alack, her art died with her. There be base and weakling imitations left, but no true blasphemy."

The Ruffler sighed; the listeners sighed in sympathy; a general depression fell upon the company for a moment, for even hardened outcasts like these are not wholly dead to sentiment, but are able to feel a fleeting sense of loss and affliction at wide intervals and under peculiarly favoring circumstances — as in cases like to this, for instance, when genius and culture depart and leave no heir. However, a deep drink all round soon restored the spirits of the mourners.

"WHILST THE FLAMES LICKED UPWARDS."

"Have any others of our friends fared hardly?" asked Hobbs.

"Some — yes. Particularly new comers — such as small husbandmen turned shiftless and hungry upon the world because their farms were taken from them to be changed to sheep ranges. They begged, and were whipped at the cart's tail, naked from the girdle up, till the blood ran; then set in the stocks to be pelted; they begged again, were whipped again, and deprived of an ear; they begged a third time — poor devils, what else could they do? — and were branded on the cheek with a red hot iron, then sold for slaves; they ran away, were hunted down, and hanged. 'Tis a brief tale, and quickly told. Others of us have fared less hardly. Stand forth, Yokel, Burns, and Hodge — show your adornments!"

These stood up and stripped away some of their rags, exposing their backs, criss-crossed with ropy old welts left by the lash; one turned up his hair and showed the place where a left ear had once

"THEY WERE WHIPPED AT THE CART'S TAIL."

been; another showed a brand upon his shoulder — the letter V — and a mutilated ear; the third said —

"I am Yokel, once a farmer and prosperous, with loving wife and kids — now am I somewhat different in estate and calling; and the wife and kids are gone; mayhap they are in heaven, mayhap in — in the other place — but the kindly God be thanked, they bide no more in *England!* My good old blameless mother strove to earn bread by nursing the sick; one of these died, the doctors knew not how, so my mother was burnt for a witch, whilst my babes looked on and wailed. English law! — up, all, with your cups! — now altogether and with a cheer! — drink to the merciful English law that delivered *her* from the English hell! Thank you, mates, one and all. I begged, from house to house — I and the wife — bearing with us the hungry kids — but it was crime to be hungry in England — so they stripped us and lashed us through three towns. Drink ye all again to the merciful English law! — for its lash drank deep of my Mary's blood and its blessed deliverance came quick. She lies there, in the potter's field, safe from all harms. And the kids — well, whilst the law lashed me from town to town, they starved. Drink lads — only a drop — a drop to the poor kids, that never did any creature harm. I begged again — begged for a crust, and got the stocks and lost an ear — see, here bides the stump; I begged again, and here is the stump of the other to keep me minded of it. And still I begged again, and was sold for a slave — here on my cheek under this stain, if I washed it off, ye might see the red S the branding-iron left there! A SLAVE! Do ye understand that word! An English SLAVE! — that is he that stands before ye. I have run from my master, and when I am found — the heavy curse of heaven fall on the law of the land that hath commanded it! — I shall hang!"[1]

A ringing voice came through the murky air —

"Thou shalt *not!* — and this day the end of that law is come!"

All turned, and saw the fantastic figure of the little king approaching hurriedly; as it emerged into the light and was clearly revealed, a general explosion of inquiries broke out:

"Who is it? *What* is it? Who art thou, manikin?"

[1] See Note 10, at end of volume.

The boy stood unconfused in the midst of all those surprised and questioning eyes, and answered with princely dignity —

"I am Edward, king of England."

A wild burst of laughter followed, partly of derision and partly of delight in the excellence of the joke. The king was stung. He said sharply —

"Ye mannerless vagrants, is this your recognition of the royal boon I have promised?"

"THOU SHALT NOT."

He said more, with angry voice and excited gesture, but it was lost in a whirlwind of laughter and mocking exclamations. "John Hobbs" made several attempts to make himself heard above the din, and at last succeeded — saying —

"Mates, he is my son, a dreamer, a fool, and stark mad — mind him not — he thinketh he *is* the king."

"I *am* the king," said Edward, turning toward him, "as thou shalt know to thy cost, in good time. Thou hast confessed a murder — thou shalt swing for it."

"*Thou'lt* betray me?—*thou?* An' I get my hands upon thee"—

"Tut-tut!" said the burly Ruffler, interposing in time to save the king, and emphasizing this service by knocking Hobbs down with his fist, "hast respect for neither Kings *nor* Rufflers? An' thou insult my presence so again, I'll hang thee up myself." Then he said to his majesty, "Thou must make no threats against thy mates, lad; and thou must guard thy tongue from saying evil of them elsewhere. *Be* king, if it please thy mad humor, but be not harmful in it. Sink the title thou hast uttered,—'tis treason; we be bad men, in some few trifling ways, but none among us is so base as to be traitor to his king; we be loving and loyal hearts, in that regard. Note if I speak truth. Now—all together: 'Long live Edward, king of England!'"

"KNOCKING HOBBS DOWN."

"LONG LIVE EDWARD, KING OF ENGLAND!"

The response came with such a thundergust from the motley crew that the crazy building vibrated to the sound. The little king's face lighted with pleasure for an instant, and he slightly inclined his head and said with grave simplicity —

"I thank you, my good people."

This unexpected result threw the company into convulsions of merriment. When something like quiet was presently come again, the Ruffler said, firmly, but with an accent of good nature —

"Drop it, boy, 'tis not wise, nor well. Humor thy fancy, if thou must, but choose some other title."

A tinker shrieked out a suggestion —

"Foo-foo the First, King of the Mooncalves!"

The title "took," at once, every throat responded, and a roaring shout went up, of —

"Long live Foo-foo the First, King of the Mooncalves!" followed by hootings, cat-calls, and peals of laughter.

"Hale him forth, and crown him!"

"Robe him!"

"Sceptre him!"

"Throne him!"

These and twenty other cries broke out at once; and almost before the poor little victim could draw a breath he was crowned with a tin basin, robed in a tattered blanket, throned upon a barrel, and sceptred with the tinker's soldering-iron. Then all flung themselves upon their knees about him and sent up a chorus of ironical wailings, and mocking supplications, whilst they swabbed their eyes with their soiled and ragged sleeves and aprons —

"Be gracious to us, O, sweet king!"

"Trample not upon thy beseeching worms, O noble majesty!"

"Pity thy slaves, and comfort them with a royal kick!"

"Cheer us and warm us with thy gracious rays, O flaming sun of sovereignty!"

"Sanctify the ground with the touch of thy foot, that we may eat the dirt and be ennobled!"

"Deign to spit upon us, O sire, that our children's children may tell of thy princely condescension, and be proud and happy forever!"

"THRONE HIM."

But the humorous tinker made the "hit" of the evening and carried off the honors. Kneeling, he pretended to kiss the king's foot, and was indignantly spurned; whereupon he went about begging for a rag to paste over the place upon his face which had been touched by the foot, saying it must be preserved from contact with the vulgar air, and that he should make his fortune by going on the highway and exposing it to view at the rate of a hundred shillings a sight. He made himself so killingly funny that he was the envy and admiration of the whole mangy rabble.

Tears of shame and indignation stood in the little monarch's eyes; and the thought in his heart was, "Had I offered them a deep wrong they could not be more cruel — yet have I proffered nought but to do them a kindness — and it is thus they use me for it!"

The Prince with the Tramps

CHAPTER XVIII.

THE PRINCE WITH THE TRAMPS.

THE troop of vagabonds turned out at early dawn, and set forward on their march. There was a lowering sky overhead, sloppy ground under foot, and a winter chill in the air. All gayety was gone from the company; some were sullen and silent, some were irritable and petulant, none were gentle-humored, all were thirsty.

The Ruffler put "Jack" in Hugo's charge, with some brief instructions, and commanded John Canty to keep away from him and let him alone; he also warned Hugo not to be too rough with the lad.

After a while the weather grew milder, and the clouds lifted somewhat. The troop ceased to shiver, and their spirits began to improve. They grew more and more cheerful, and finally began to chaff each other and insult passengers along the highway. This showed that they were awaking to an appreciation of life and its joys once more. The dread in which their sort was held was apparent in the fact that everybody gave them the road, and took their ribald insolences meekly, without venturing to talk back. They snatched linen from the hedges, occasionally, in full view of the owners, who made no protest, but only seemed grateful that they did not take the hedges, too.

By and by they invaded a small farm house and made themselves at home while the trembling farmer and his people swept the larder clean to furnish a breakfast for them. They chucked the housewife and her daughters under the chin whilst receiving the food from their hands, and made coarse jests about them, accompanied with insulting epithets and bursts of horse-laughter. They threw bones and vegetables at the farmer and his sons, kept them dodging all the time, and applauded uproariously when a good hit was made. They ended by buttering the head of one of the daughters who resented some of their familiarities. When they took their leave they threatened to come back and burn the house over the heads of the family if any report of their doings got to the ears of the authorities.

"THE TROOP OF VAGABONDS SET FORWARD."

About noon, after a long and weary tramp, the gang came to a halt behind a hedge on the outskirts of a considerable village. An hour was allowed for rest, then the crew scattered themselves abroad to enter the village at different points to ply their various trades.—"Jack" was sent with Hugo. They wandered hither and thither for some time, Hugo watching for opportunities to do a stroke of business but finding none — so he finally said —

"I see nought to steal; it is a paltry place. Wherefore we will beg."

"*We*, forsooth! Follow thy trade — it befits thee. But *I* will not beg."

"Thou'lt not beg!" exclaimed Hugo, eying the king with surprise. "Prithee, since when hast thou reformed?"

"What dost thou mean?"

"Mean? Hast thou not begged the streets of London all thy life?"

"I? Thou idiot!"

"THEY THREW BONES AND VEGETABLES."

"Spare thy compliments — thy stock will last the longer. Thy father says thou hast begged all thy days. Mayhap he lied. Peradventure you will even make so bold as to *say* he lied," scoffed Hugo.

"Him *you* call my father? Yes, he lied."

"Come, play not thy merry game of madman so far, mate; use it for thy amusement, not thy hurt. An' I tell him this, he will scorch thee finely for it."

"Save thyself the trouble. I will tell him."

"I like thy spirit, I do in truth; but I do not admire thy judgment. Bone-rackings and bastings be plenty enow in this life, without going out of one's way to invite them. But a truce to these matters; *I* believe your father. I doubt not he can lie; I doubt not he *doth* lie, upon occasion, for the best of us do that; but there is no occasion here. A wise man does not waste so good a commodity as lying for nought. But come; sith it is thy humor to give over begging, wherewithal shall we busy ourselves? With robbing kitchens?"

The king said, impatiently —

"Have done with this folly — you weary me!"

Hugo replied, with temper —

"Now harkee, mate; you will not beg, you will not rob; so be it. But I will tell you what you *will* do. You will play decoy whilst *I* beg. Refuse, an' you think you may venture!"

The king was about to reply contemptuously, when Hugo said, interrupting —

"Peace! Here comes one with a kindly face. Now will I fall down in a fit. When the stranger runs to me, set you up a wail, and fall upon your knees, seeming to weep; then cry out as all the devils of misery were in your belly, and say, 'O, sir, it is my poor afflicted brother, and we be friendless; o' God's name cast through your merciful eyes one pitiful look upon a sick, forsaken and most miserable wretch; bestow one little penny out of thy riches upon one smitten of God and ready to perish!' — and mind you, keep you *on* wailing, and abate not till we bilk him of his penny, else shall you rue it."

Then immediately Hugo began to moan, and groan, and roll his eyes, and reel and totter about; and when the stranger was close at hand, down he sprawled before him, with a shriek, and began to writhe and wallow in the dirt, in seeming agony.

"O dear, O dear!" cried the benevolent stranger, "O poor soul, poor soul, how he doth suffer! There — let me help thee up."

"O, noble sir, forbear, and God love you for a princely gentleman — but it giveth me cruel pain to touch me when I am taken so. My brother there will tell your worship how I am racked with anguish when these fits be upon me. A penny, dear sir, a penny, to buy a little food; then leave me to my sorrows."

"A penny! thou shalt have three, thou hapless creature" — and he fumbled in his pocket with nervous haste and got them out. "There, poor lad, take

"BEGAN TO WRITHE AND WALLOW IN THE DIRT."

them, and most welcome. Now come hither, my boy, and help me carry thy stricken brother to yon house, where" —

"I am not his brother," said the king, interrupting.

"What! not his brother?"

"O hear him!" groaned Hugo, then privately ground his teeth. "He denies his own brother — and he with one foot in the grave!"

"Boy, thou art indeed hard of heart, if this is thy brother. For shame! — and he scarce able to move hand or foot. If he is not thy brother, who is he, then?"

"A beggar and a thief! He has got your money and has picked your pocket likewise. An' thou wouldst do a healing miracle, lay thy staff over his shoulders and trust Providence for the rest."

"THE KING FLED IN THE OPPOSITE DIRECTION."

But Hugo did not tarry for the miracle. In a moment he was up and off like the wind, the gentleman following after and raising the hue and cry lustily as he went. The king, breathing deep gratitude to Heaven for his own release, fled in the opposite direction and did not slacken his pace until he was out of harm's reach. He took the first road that offered, and soon put the village behind him. He hurried along, as briskly as he could, during several hours, keeping a nervous watch over his shoulder for pursuit; but his fears left him at last, and a grateful sense of security took their place. He recognized, now, that he was hungry; and also very tired. So he halted at a farm house; but when he was about to speak, he was cut short and driven rudely away. His clothes were against him.

He wandered on, wounded and indignant, and was resolved to put himself in the way of like treatment no more. But hunger is pride's master; so as the evening drew near, he made an attempt at another farm house; but here he fared worse than before; for he was called hard names and was promised arrest as a vagrant except he moved on promptly.

The night came on, chilly and overcast; and still the footsore monarch labored slowly on. He was obliged to keep moving, for every time he sat down to rest he was soon penetrated to the bone with the cold. All his sensations and experiences, as he moved through the solemn gloom and the empty vastness of the night, were new and strange to him. At intervals he heard voices approach, pass by, and fade into silence; and as he saw nothing more of the bodies they belonged to than a sort of formless drifting blur, there was something spectral and uncanny about it all that made him shudder. Occasionally he caught the twinkle of a light — always far away, apparently — almost in another world; if he heard the tinkle of a sheep's bell, it was vague, distant, indistinct; the muffled lowing of the herds floated to him on the night wind in vanishing cadences, a mournful sound; now and then came the complaining howl of a dog over viewless expanses of field and forest; all sounds were remote; they made the little king feel that all life and activity were far removed from him, and that he stood solitary, companionless, in the centre of a measureless solitude.

He stumbled along, through the grewsome fascinations of this new experience, startled occasionally by the soft rustling of the dry leaves overhead, so like human whispers they seemed to sound; and by and by he came suddenly upon the freckled light of a tin lantern near at hand. He stepped back into the shadows and waited. The lantern stood by the open door of a barn. The king waited some time — there was no sound, and nobody stirring. He got so cold, standing still, and the hospitable barn looked so enticing, that at last he resolved to risk every thing and enter. He started swiftly and stealthily, and just as he was crossing the threshold he heard voices behind him. He darted behind a cask, within the barn, and stooped down. Two

"HE STUMBLED ALONG."

farm laborers came in, bringing the lantern with them, and fell to work, talking meanwhile. Whilst they moved about with the light, the king made good use of his eyes and took the bearings of what seemed to be a good sized stall at the further end of the place, purposing to grope his way to it when he should be left to himself. He also noted the position of a pile of horse blankets, midway of the route, with the intent to levy upon them for the service of the crown of England for one night.

By and by the men finished and went away, fastening the door behind them and taking the lantern with them. The shivering king made for the blankets, with as good speed as the darkness would allow; gathered them up and then groped his way safely to the stall. Of two of the blankets he made a bed, then covered himself with the remaining two. He was a glad monarch, now, though the blankets were old and thin, and not quite warm enough; and besides gave out a pungent horsy odor that was almost suffocatingly powerful.

Although the king was hungry and chilly, he was also so tired and so drowsy that these latter influences soon began to get the advantage of the former, and he presently dozed off into a state of semi-consciousness. Then, just as he was on the point of losing himself wholly, he distinctly felt something touch him! He was broad awake in a moment, and gasping for breath. The cold horror of that mysterious touch in the dark almost made his heart stand still. He lay motionless, and listened, scarcely breathing. But nothing stirred, and there was no sound. He continued to listen, and wait, during what seemed a long time, but still nothing stirred, and there was no sound. So he began to drop into a drowse once more, at last; and all at once he felt that mysterious touch again! It was a grisly thing, this light touch from this noiseless and invisible presence; it made the boy sick with ghostly fears. What should he do? That was the question; but he did not know how to answer it. Should he leave these reasonably comfortable quarters and fly from this inscrutable horror? But fly whither? He could not get out of the barn; and the idea of scurrying blindly hither and thither in the dark, within the captivity of the four walls, with this phantom gliding after him, and visiting him with that soft hideous touch upon cheek or shoulder at every turn, was intolerable. But to stay where he was, and endure this living death all night?—was that better? No. What, then, was there left to do? Ah, there was but one course; he knew it well—he must put out his hand and find that thing!

It was easy to think this; but it was hard to brace himself up to try it. Three times he stretched his hand a little way out into the dark, gingerly; and snatched it suddenly back, with a gasp — not because it had encountered any thing, but because he had felt so sure it was just *going* to. But the fourth time, he groped a little further, and his hand lightly swept against something soft and warm. This

"WHAT SEEMED TO BE A WARM ROPE."

petrified him, nearly, with fright — his mind was in such a state that he could imagine the thing to be nothing else than a corpse, newly dead and still warm. He thought he would rather die than touch it again. But he thought this false thought because he did not know the immortal strength of human curiosity. In no long time his hand was tremblingly groping again — against his judgment, and without his consent — but groping persistently on, just the same. It encountered a bunch of long hair; he shuddered, but followed up the hair and found what seemed to be a warm rope; followed up the rope and found an innocent calf! — for the rope was not a rope at all, but the calf's tail.

The king was cordially ashamed of himself for having gotten all that fright and misery out of so paltry a matter as a slumbering calf; but he need not have felt so about it, for it was not the calf that frightened him but a dreadful non-existent something which the calf stood for; and any other boy, in those old superstitious times, would have acted and suffered just as he had done.

The king was not only delighted to find that the creature was only a calf, but delighted to have the calf's company; for he had been feeling so lonesome and friendless that the company and comradeship of even this humble animal was welcome. And he had been so buffeted, so rudely entreated by his own kind, that it was a real comfort to him to feel that he was at last in the society of a fellow creature that had at least a soft heart and a gentle spirit, whatever loftier attributes might be lacking. So he resolved to waive rank and make friends with the calf.

"CUDDLED UP TO THE CALF."

While stroking its sleek warm back — for it lay near him and within easy reach — it occurred to him that this calf might be utilized in more ways than one. Whereupon he re-arranged his bed, spreading it down close to the calf; then he cuddled himself up to the calf's back, drew the covers up over himself and his friend, and in a minute or two was as warm and comfortable as he had ever been in the downy couches of the regal palace of Westminster.

Pleasant thoughts came, at once; life took on a cheerfuller seeming. He was free of the bonds of servitude and crime, free of the companionship of base and brutal outlaws; he was warm, he was sheltered; in a word, he was happy. The night wind was rising; it swept by in fitful gusts that made the old barn quake and rattle, then its forces died down at intervals, and went moaning and wailing around corners and projections — but it was all music to the king, now that he was snug and comfortable: let it blow and rage, let it batter and bang, let it moan and wail, he minded it not, he only enjoyed it. He merely snuggled the closer to his friend, in a luxury of warm contentment, and drifted blissfully out of consciousness into a deep and dreamless sleep that was full of serenity and peace. The distant dogs howled, the melancholy kine complained, and the winds went on raging, whilst furious sheets of rain drove along the roof; but the majesty of England slept on, undisturbed, and the calf did the same, it being a simple creature and not easily troubled by storms or embarrassed by sleeping with a king.

THE PRINCE
WITH THE
PEASANTS

CHAPTER XIX.

THE PRINCE WITH THE PEASANTS.

When the king awoke in the early morning, he found that a wet but thoughtful rat had crept into the place during the night and made a cosey bed for itself in his bosom. Being disturbed, now, it scampered away. The boy smiled, and said, "Poor fool, why so fearful? I am as forlorn as thou. 'Twould be a shame in me to hurt the helpless, who am myself so helpless. Moreover, I owe you thanks for a good omen; for when a king has fallen so low that the very rats do make a bed of him, it surely meaneth that his fortunes be upon the turn, since it is plain he can no lower go."

He got up and stepped out of the stall, and just then he heard the sound of children's voices. The barn door opened and a couple of little girls came in. As soon as they saw him their talking and laughing ceased, and they stopped and stood still, gazing at him with strong curiosity; they presently began to whisper together, then they approached nearer, and stopped again to gaze and whisper. By and by they gathered courage and began to discuss him aloud. One said —

"He hath a comely face."

The other added —

"And pretty hair."

"But is ill clothed enow."

"And how starved he looketh."

They came still nearer, sidling shyly around and about him, examining him minutely from all points, as if he were some strange new kind of animal; but warily and watchfully, the while, as if they half feared he might be a sort of animal that would bite, upon occasion. Finally they halted before him, holding each other's hands, for protection, and took a good satisfying stare with their innocent eyes; then one of them plucked up all her courage and inquired with honest directness —

"Who art thou, boy?"

"I am the king," was the grave answer.

The children gave a little start, and their eyes spread themselves wide open and remained so during a speechless half minute. Then curiosity broke the silence —

"The *king?* What king?"

"The king of England."

The children looked at each other — then at him — then at each other again — wonderingly, perplexedly — then one said —

"Didst hear him, Margery? — he saith he is the king. Can that be true?"

"How can it be else but true, Prissy? Would he say a lie? For look you, Prissy, an' it were not true, it *would* be a lie. It surely would be. Now think on't. For all things that be not true, be lies — thou canst make nought else out of it."

It was a good tight argument, without a leak in it anywhere; and it left Prissy's half-doubts not a leg to stand on. She considered a moment, then put the king upon his honor with the simple remark —

"If thou art truly the king, then I believe thee."

"I am truly the king."

This settled the matter. His majesty's royalty was accepted without further question or discussion, and the two little girls began at once to inquire into how he came to be where he was, and how he came to be so unroyally clad, and whither he was bound, and all about his affairs. It was a mighty relief to him to pour out his troubles where they would not be scoffed at or doubted; so he told his tale with feeling, forgetting even his hunger for the time; and it was received with the deepest and tenderest sympathy by the gentle little maids. But when he got down to his latest experiences and they learned how long he had been without food, they cut him short and hurried him away to the farm house to find a breakfast for him.

"TOOK A GOOD SATISFYING STARE."

The king was cheerful and happy, now, and said to himself, "When I am come to mine own again, I will always honor little children, remembering how that these trusted me and believed in me in my time of trouble; whilst they that were older, and thought themselves wiser, mocked at me and held me for a liar."

The children's mother received the king kindly, and was full of pity; for his forlorn condition and apparently crazed intellect touched her womanly heart. She was a widow, and rather poor; consequently she had seen trouble enough to enable her to feel for the unfortunate. She imagined that the demented boy had wandered away from his friends or keepers; so she tried to find out whence he had come, in order that she might take measures to return him; but all her references to neighboring towns and villages, and all her inquiries in the same line, went for nothing — the boy's face, and his answers, too, showed that the things she was talking of were not familiar to him. He spoke earnestly and simply about court matters; and broke down,

"THE CHILDREN'S MOTHER RECEIVED THE KING KINDLY."

more than once, when speaking of the late king "his father;" but whenever the conversation changed to baser topics, he lost interest and became silent.

The woman was mightily puzzled; but she did not give up. As she proceeded with her cooking, she set herself to contriving devices to surprise the boy into betraying his real secret. She talked about cattle — he showed no concern; then about sheep — the same result — so her guess that he had been a shepherd boy was an error; she talked about mills; and about weavers, tinkers, smiths, trades and tradesmen of all sorts; and about Bedlam, and jails, and charitable retreats; but no matter, she was baffled at all points. Not altogether, either; for she argued that she had narrowed the thing down to domestic service. Yes, she was sure she was on the right track, now — he must have been a house servant. So she led up to that. But the result was discouraging. The subject of sweeping appeared to weary him; fire-building failed to stir him; scrubbing and scouring awoke no enthusiasm. Then the goodwife touched, with a perishing hope, and rather as a matter of form, upon the subject of cooking. To her surprise, and her vast delight, the king's face lighted at once! Ah, she had hunted him down at last, she thought; and she was right proud too, of the devious shrewdness and tact which had accomplished it.

Her tired tongue got a chance to rest, now; for the king's, inspired by gnawing hunger and the fragrant smells that came from the sputtering pots and pans, turned itself loose and delivered itself up to such an eloquent dissertation upon certain toothsome dishes, that within three minutes the woman said to herself, "Of a truth I was right — he hath holpen in a kitchen!" Then he broadened his bill of fare, and discussed it with such appreciation and animation, that the goodwife said to herself, "Good lack! how can he know so many dishes, and so fine ones withal? For these belong only upon the tables of the rich and great. Ah, now I see! ragged outcast as he is, he must have served in the palace before his reason went astray; yes, he must have helped in the very kitchen of the king himself! I will test him."

Full of eagerness to prove her sagacity, she told the king to mind the cooking a moment — hinting that he might manufacture and add a dish or two, if he chose — then she went out of the room and gave her children a sign to follow after. The king muttered —

"Another English king had a commission like to this, in a bygone time — it is nothing against my dignity to undertake an office which the great Alfred stooped to assume. But I will try to better serve my trust than he; for he let the cakes burn."

The intent was good, but the performance was not answerable to it; for this king, like the other one, soon fell into deep thinkings concerning his vast affairs, and the same calamity resulted — the cookery got burned. The woman returned in time to save the breakfast from entire destruction; and she promptly brought the king out of his dreams with a brisk and cordial tongue-lashing. Then, seeing how troubled he was, over his violated trust, she softened at once and was all goodness and gentleness toward him.

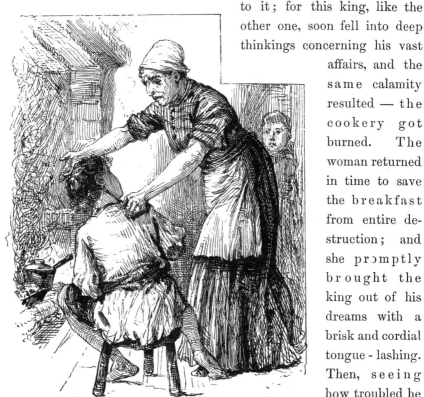

"BROUGHT THE KING OUT OF HIS DREAMS."

The boy made a hearty and satisfying meal, and was greatly refreshed and gladdened by it. It was a meal which was distinguished by this curious feature, that rank was waived on both sides; yet neither recipient of the favor was aware that it had been extended. The goodwife had intended to feed this young tramp with broken victuals in a corner, like any other tramp, or like a dog; but she was so remorseful for the scolding she had given him, that she did what she could to atone for it by allowing him to sit at the family table and eat with his betters, on ostensible terms of equality with them; and the king, on his side, was so remorseful for having broken his trust, after the family had been so kind to him, that he forced himself to atone for it by humbling himself to the family level, instead of requiring the woman and her children to stand and wait upon him while he occupied their table in the solitary state due his birth and dignity. It does us all good to unbend sometimes. This good woman was made happy all the day long by the applauses which she got out of herself for her magnanimous condescension to a tramp; and the king was just as self-complacent over his gracious humility toward a humble peasant woman.

When breakfast was over, the housewife told the king to wash up the dishes. This command was a staggerer, for a moment, and the king came near rebelling; but then he said to himself, "Alfred the Great watched the cakes; doubtless he would have washed the dishes, too—therefore will I essay it."

He made a sufficiently poor job of it; and to his surprise, too, for the cleaning of wooden spoons and trenchers had seemed an easy thing to do. It was a tedious and troublesome piece of work, but he finished it at last. He was becoming impatient to get away on his journey now; however, he was not to lose this thrifty dame's society so easily. She furnished him some little odds and ends of employment, which he got through with after a fair fashion and with some credit. Then she set him and the little girls to paring some winter apples; but he was so awkward at this service, that she retired him from it and gave him a butcher knife to grind. Afterward she kept him carding wool until he began to think he had laid the good King Alfred about far enough in the shade for the present, in the matter of showy menial heroisms that would read picturesquely in story-books and histories, and so he was half minded to resign. And when, just after the noonday dinner, the goodwife gave him a basket of kittens to drown, he did resign. At least he was just going to resign — for he felt that he must draw the line somewhere, and it seemed to him that to draw it at kitten-drowning was about the right thing — when there was an interruption. The interruption was John Canty — with a peddler's pack on his back — and Hugo!

"GAVE HIM A BUTCHER KNIFE TO GRIND."

The King discovered these rascals approaching the front gate before they had had a chance to see him; so he said nothing about drawing the line, but took up his basket of kittens and stepped quietly out the back way, without a word. He left the creatures in an outhouse, and hurried on, into a narrow lane at the rear.

THE PRINCE
AND
THE HERMIT

CHAPTER XX.

THE PRINCE AND THE HERMIT.

THE high hedge hid him from the house, now; and so, under the impulse of a deadly fright, he let out all his forces and sped toward a wood in the distance. He never looked back until he had almost gained the shelter of the forest; then he turned and descried two figures in the distance. That was sufficient; he did not wait to scan them critically, but hurried on, and never abated his pace till he was far within the twilight depths of the wood. Then he stopped; being persuaded that he was now tolerably safe. He listened intently, but the stillness was profound and solemn — awful, even, and depressing to the spirits. At wide intervals his straining ear did detect sounds, but they were so remote, and hollow, and mysterious, that they seemed not to be real sounds, but only the moaning and complaining ghosts of departed ones. So the sounds were yet more dreary than the silence which they interrupted.

It was his purpose, in the beginning, to stay where he was, the rest of the day; but a chill soon invaded his perspiring body, and he was at last obliged to resume movement in order to get warm. He struck straight through the forest, hoping to pierce to a road presently, but he was disappointed in this. He travelled on and on; but the farther he went, the denser the wood became, apparently. The gloom began to thicken, by and by, and the king realized that the night was coming on. It made him shudder to think of spending it in such an uncanny place; so he tried to hurry faster, but he only made the less speed, for he could not now see well enough to choose his steps judiciously; consequently he kept tripping over roots and tangling himself in vines and briers.

And how glad he was when at last he caught the glimmer of a light! He approached it warily, stopping often to look about him and listen. It came

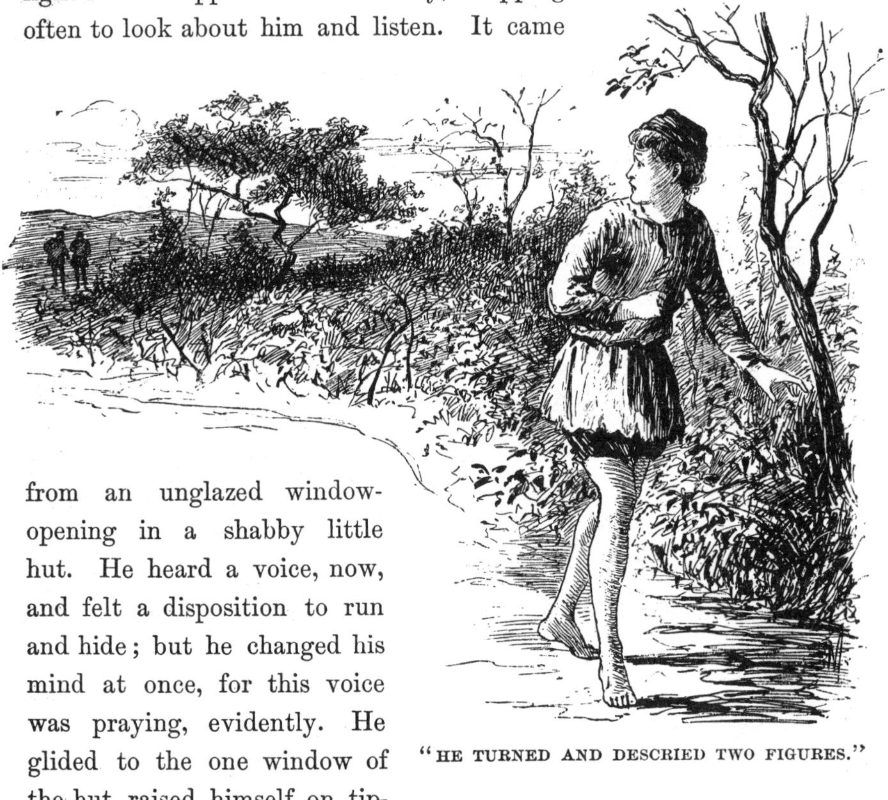

"HE TURNED AND DESCRIED TWO FIGURES."

from an unglazed window-opening in a shabby little hut. He heard a voice, now, and felt a disposition to run and hide; but he changed his mind at once, for this voice was praying, evidently. He glided to the one window of the hut, raised himself on tiptoe, and stole a glance within. The room was small; its floor was the natural earth, beaten hard by use; in a corner was a bed of rushes and a ragged blanket or two; near it was a pail, a cup, a basin, and two or three pots and pans; there was a short bench and a three-legged stool;

on the hearth the remains of a fagot fire were smouldering; before a shrine, which was lighted by a single candle, knelt an aged man, and on an old wooden box at his side, lay an open book and a human skull. The man was of large, bony frame; his hair and whiskers were very

"THE KING ENTERED AND PAUSED."

long and snowy white; he was clothed in a robe of sheepskins which reached from his neck to his heels.

"A holy hermit!" said the king to himself; "now am I indeed fortunate."

The hermit rose from his knees; the king knocked. A deep voice responded —

"Enter! — but leave sin behind, for the ground whereon thou shalt stand is holy!"

The king entered, and paused. The hermit turned a pair of gleaming, unrestful eyes upon him, and said —

"Who art thou?"

"I am the king," came the answer, with placid simplicity.

"Welcome, king!" cried the hermit, with enthusiasm. Then, bustling about with feverish activity, and constantly saying "Welcome, welcome," he arranged his bench, seated the king on it, by the hearth, threw some fagots on the fire, and finally fell to pacing the floor, with a nervous stride.

"Welcome! Many have sought sanctuary here, but they were not worthy, and were turned away. But a king who casts his crown away, and despises the vain splendors of his office, and clothes his body in rags, to devote his life to holiness and the mortification of the flesh — he is worthy, he is welcome! — here shall he abide all his days till death come." The king hastened to interrupt and explain, but the hermit paid no attention to him — did not even hear him, apparently, but went right on with his talk, with a raised voice and a growing energy. "And thou shalt be at peace here. None shall find out thy refuge to disquiet thee with supplications to return to that empty and foolish life which God hath moved thee to abandon. Thou shalt pray, here; thou shalt study the Book; thou shalt meditate upon the follies and delusions of this world, and upon the sublimities of the world to come; thou shalt feed upon crusts and herbs, and scourge thy body with whips, daily, to the purifying of thy soul. Thou shalt wear a hair shirt next thy skin; thou shalt drink water, only; and thou shalt be at peace; yes, wholly at peace; for whoso comes to seek thee shall go his way again, baffled; he shall not find thee, he shall not molest thee."

The old man, still pacing back and forth, ceased to speak aloud, and began to mutter. The king seized this opportunity to state his case; and he did it with an eloquence inspired by uneasiness and apprehension. But the hermit went on muttering, and gave no heed. And still muttering, he approached the king and said, impressively —

"'Sh! I will tell you a secret!" He bent down to impart it, but checked himself, and assumed a listening attitude. After a moment or two he went on tiptoe to the window-opening, put his head out and peered around in the gloaming, then came tiptoeing back again, put his face close down to the king's, and whispered —

"I am an archangel!"

The king started violently, and said to himself, "Would God I were with the outlaws again; for lo, now am I the prisoner of a madman!" His apprehensions were heightened, and they showed plainly in his face. In a low, excited voice, the hermit continued —

"I WILL TELL YOU A SECRET."

"I see you feel my atmosphere! There's awe in your face! None may be in this atmosphere and not be thus affected; for it is the very atmosphere of heaven. I go thither and return, in the twinkling of an eye. I was made an archangel on this very spot, it is five years ago, by angels sent from heaven to confer that awful dignity. Their presence filled this place with an intolerable brightness. And they knelt to me, king! yes, they knelt to me! for I was greater than they. I have walked in the courts of heaven, and held speech with the patriarchs. Touch my hand — be not afraid — touch it. There — now thou hast touched a hand which has been clasped by Abraham, and Isaac and Jacob! For I have walked in the golden courts, I have seen the Deity face to face!" He paused, to give this speech effect; then his face suddenly changed, and he started to his feet again, saying, with angry energy, "Yes, I am an archangel; *a mere archangel!* — I that might have been pope! It is verily true. I was told it from heaven in a dream, twenty years ago; ah, yes, I was to be pope! — and I *should* have been pope, for Heaven had said it — but the king dissolved my religious house, and I, poor obscure unfriended monk, was cast homeless upon the world, robbed of my mighty destiny!" Here he began to mumble again, and beat his forehead in futile rage, with his fist; now and then articulating a venomous curse, and now and then a pathetic "Wherefore I am nought but an archangel — I that should have been pope!"

So he went on, for an hour, whilst the poor little king sat and suffered. Then all at once the old man's frenzy departed, and he became all gentleness. His voice softened, he came down out of his clouds, and fell to prattling along so simply and so humanly, that he soon won the king's heart completely. The old devotee moved the boy nearer to the fire and made him comfortable; doctored his small bruises and abrasions with a deft and tender hand; and then set about preparing and cooking a supper — chatting pleasantly all the time, and occasionally stroking the lad's cheek or patting his head, in such a gently caressing way that in a little while all the fear and repulsion inspired by the archangel were changed to reverence and affection for the man.

This happy state of things continued while the two ate the supper; then, after a prayer before the shrine, the hermit put the boy to bed, in a small adjoining room, tucking him in as snugly and lovingly as a mother might; and so, with a parting caress, left him and sat down by the fire, and began to poke the brands about in an absent and aimless way. Presently he paused; then tapped his forehead several times with his fingers, as if trying to recall some thought which had escaped from his mind. Apparently he was unsuccessful. Now he started quickly up, and entered his guest's room, and said—

"CHATTING PLEASANTLY ALL THE TIME."

"Thou art king?"

"Yes," was the response, drowsily uttered.

"What king?"

"Of England."

"Of England! Then Henry is gone!"

"Alack, it is so. I am his son."

A black frown settled down upon the hermit's face, and he clenched his bony hands with a vindictive energy. He stood a few moments, breathing fast and swallowing repeatedly, then said in a husky voice—

"Dost know it was he that turned us out into the world houseless and homeless?"

There was no response. The old man bent down and scanned the boy's reposeful face and listened to his placid breathing. "He sleeps — sleeps soundly;" and the frown vanished away and gave place to an expression of evil satisfaction. A smile flitted across the dreaming boy's features. The hermit muttered, "So — his heart is happy;" and he turned away. He went stealthily about the place, seeking here and there for something; now and then halting to listen, now and then jerking his head around and casting a quick glance toward the bed; and always muttering, always mumbling to himself. At last he found what he seemed to want — a rusty old butcher knife and a whetstone. Then he crept to his place by the fire, sat himself down, and began to whet the knife softly on the stone, still muttering, mumbling, ejaculating. The winds sighed around the lonely place, the mysterious voices of the night floated by out of the distances. The shining eyes of venturesome mice and rats peered out at the old man from cracks and coverts, but he went on with his work, rapt absorbed, and noted none of these things.

At long intervals he drew his thumb along the edge of his knife, and nodded his head with satisfaction. "It grows sharper," he said; "yes, it grows sharper."

He took no note of the flight of time, but worked tranquilly on, entertaining himself with his thoughts, which broke out occasionally in articulate speech:

"His father wrought us evil, he destroyed us — and is gone down into the eternal fires! Yes, down into the eternal fires! He escaped us — but it was God's will, yes it was God's will, we must not repine. But he hath not escaped the fires! no, he hath not escaped the fires, the consuming, unpitying, remorseless fires — and *they* are everlasting!"

And so he wrought; and still wrought; mumbling—chuckling a low rasping chuckle, at times—and at times breaking again into words:

"It was his father that did it all. I am but an archangel—but for him, I should be pope!"

The king stirred. The hermit sprang noiselessly to the bedside, and went down upon his knees, bending over the prostrate form with his knife uplifted. The boy stirred again; his eyes came open for an instant, but there was no speculation in them, they saw nothing; the next moment his tranquil breathing showed that his sleep was sound once more.

"DREW HIS THUMB ALONG THE EDGE."

The hermit watched and listened, for a time, keeping his position and scarcely breathing; then he slowly lowered his arm, and presently crept away, saying,—

"It is long past midnight — it is not best that he should cry out, lest by accident some one be passing."

He glided about his hovel, gathering a rag here, a thong there, and another one yonder; then he returned, and by careful and gentle handling, he managed to tie the king's ankles together without wak-

"THE NEXT MOMENT THEY WERE BOUND."

ing him. Next he essayed to tie the wrists; he made several attempts to cross them, but the boy always drew one hand or the other away, just as the cord was ready to be applied; but at last, when the archangel was almost ready to despair, the boy crossed his hands himself, and the next moment they were bound. Now a bandage was passed under the sleeper's chin and brought up over his head and tied fast — and so softly, so gradually, and so deftly were the knots drawn together and compacted, that the boy slept peacefully through it all without stirring.

Hendon to the Rescue

CHAPTER XXI.

HENDON TO THE RESCUE.

THE old man glided away, stooping, stealthy, cat-like, and brought the low bench. He seated himself upon it, half his body in the dim and flickering light, and the other half in shadow; and so, with his craving eyes bent upon the slumbering boy, he kept his patient vigil there, heedless of the drift of time, and softly whetted his knife, and mumbled and chuckled; and in aspect and attitude he resembled nothing so much as a grizzly, monstrous spider, gloating over some hapless insect that lay bound and helpless in his web.

After a long while, the old man, who was still gazing, — yet not seeing, his mind having settled into a dreamy abstraction, — observed on a sudden, that the boy's eyes were open — wide open and staring! — staring up in frozen horror at the knife. The smile of a gratified devil crept over the old man's face, and he said, without changing his attitude or his occupation —

"Son of Henry the Eighth, hast thou prayed?"

The boy struggled helplessly in his bonds; and at the same time forced a smothered sound through his closed jaws, which the hermit chose to interpret as an affirmative answer to his question.

"Then pray again. Pray the prayer for the dying!"

A shudder shook the boy's frame, and his face blenched. Then he struggled again to free himself — turning and twisting himself this way and that; tugging frantically, fiercely, desperately — but uselessly

— to burst his fetters: and all the while the old ogre smiled down upon him, and nodded his head, and placidly whetted his knife; mumbling, from time to time. "The moments are precious, they are few and precious — pray the prayer for the dying!"

The boy uttered a despairing groan, and ceased from his struggles, panting. The tears came, then, and trickled, one after the other, down his face; but this piteous sight wrought no softening effect upon the savage old man.

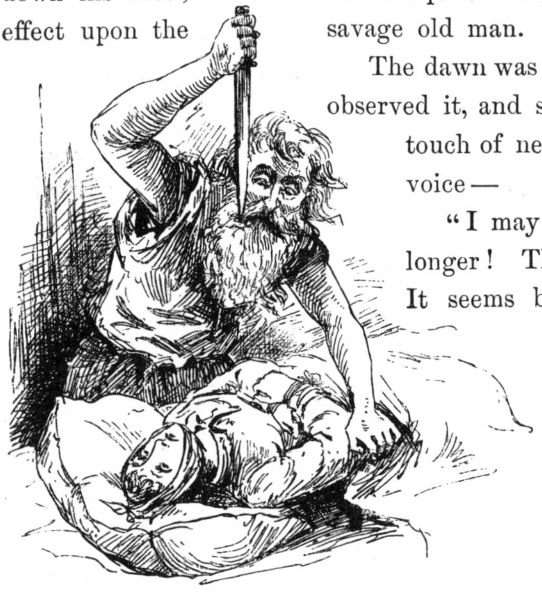

"HE SUNK UPON HIS KNEES, HIS KNIFE IN HAND."

The dawn was coming, now; the hermit observed it, and spoke up sharply, with a touch of nervous apprehension in his voice —

"I may not indulge this ecstasy longer! The night is already gone. It seems but a moment — only a moment; would it had endured a year! Seed of the Church's spoiler, close thy perishing eyes, an' thou fearest to look upon" . . .

The rest was lost in inarticulate mutterings. The old man sunk upon his knees, his knife in his hand, and bent himself over the moaning boy —

Hark! There was a sound of voices near the cabin — the knife dropped from the hermit's hand; he cast a sheepskin over the boy and started up, trembling. The sounds increased, and presently the voices became rough and angry; then came blows, and cries for help; then a clatter of swift footsteps, retreating. Immediately came a succession of thundering knocks upon the cabin door, followed by —

"Hullo-o-o! Open! And despatch, in the name of all the devils!"

O, this was the blessedest sound that had ever made music in the king's ears; for it was Miles Hendon's voice!

The hermit, grinding his teeth in impotent rage, moved swiftly out of the bedchamber, closing the door behind him; and straightway the king heard a talk, to this effect, proceeding from the "chapel:"

"Homage and greeting, reverend sir! Where is the boy — *my* boy?"

"What boy, friend?"

"What boy! Lie me no lies, sir priest, play me no deceptions! — I am not in the humor for it. Near to this place I caught the scoundrels who I judged did steal him from me, and I made them confess; they said he was at large again, and they had tracked him to your door. They showed me his very footprints. Now palter no more; for look you, holy sir, an' thou produce him not — Where is the boy?"

"O, good sir, peradventure you mean the ragged regal vagrant that tarried here the night. If such as you take interest in such as he, know, then, that I have sent him of an errand. He will be back anon."

"How soon? How soon? Come, waste not the time — cannot I overtake him? How soon will he be back?"

"Thou needst not stir; he will return quickly."

"So be it then. I will try to wait. But stop! — *you* sent him of an errand? — you! Verily this is a lie — he would not go. He would pull thy old beard, an' thou didst offer him such an insolence. Thou hast lied, friend; thou hast surely lied! He would not go for thee nor for any man."

"For any *man* — no; haply not. But I am not a man."

"*What!* Now o' God's name what art thou, then?"

"It is a secret — mark thou reveal it not. I am an archangel!"

There was a tremendous ejaculation from Miles Hendon — not altogether unprofane — followed by —

"This doth well and truly account for his complaisance! Right well I knew he would budge nor hand nor foot in the menial service of any mortal; but lord, even a king must obey when an archangel gives the word o' command! Let me — 'sh! What noise was that?"

All this while the little king had been

"GOD MADE EVERY CREATURE BUT YOU!"

yonder, alternately quaking with terror and trembling with hope; and all the while, too, he had thrown all the strength he could into his anguished moanings, constantly expecting them to reach Hendon's ear, but always realizing, with bitterness, that they failed, or at least made no impression. So this last remark of his servant came as comes a reviving breath from fresh fields to the dying; and he exerted himself once more, and with all his energy, just as the hermit was saying —

"Noise? I heard only the wind."

"Mayhap it was. Yes, doubtless that was it. I have been hearing it faintly all the—there it is again! It is not the wind! What an odd sound! Come, we will hunt it out!"

Now the king's joy was nearly insupportable. His tired lungs did their utmost—and hopefully, too—but the sealed jaws and the muffling sheepskin sadly crippled the effort. Then the poor fellow's heart sank, to hear the hermit say—

"Ah, it came from without—I think from the copse yonder. Come, I will lead the way."

The king heard the two pass out, talking; heard their footsteps die quickly away—then he was alone with a boding, brooding, awful silence.

It seemed an age till he heard the steps and voices approaching again—and this time he heard an added sound,—the trampling of hoofs, apparently. Then he heard Hendon say—

"I will not wait longer. I *cannot* wait longer. He has lost his way in this thick wood. Which direction took he? Quick—point it out to me."

"He—but wait; I will go with thee."

"Good—good! Why, truly thou art better than thy looks. Marry I do think there's not another archangel with so right a heart as thine. Wilt ride? Wilt take the wee donkey that's for my boy, or wilt thou fork thy holy legs over this ill-conditioned slave of a mule that I have provided for myself?—and had been cheated in, too, had he cost but the indifferent sum of a month's usury on a brass farthing let to a tinker out of work."

"No—ride thy mule, and lead thine ass; I am surer on mine own feet, and will walk."

"Then prithee mind the little beast for me while I take my life in my hands and make what success I may toward mounting the big one."

Then followed a confusion of kicks, cuffs, tramplings and plungings, accompanied by a thunderous intermingling of volleyed curses, and finally a bitter apostrophe to the mule, which must have broken its spirit, for hostilities seemed to cease from that moment.

With unutterable misery the fettered little king heard the voices and footsteps fade away and die out. All hope forsook him, now, for the moment, and a dull despair settled down upon his heart. "My only friend is deceived and got rid of," he said; "the

"THE FETTERED LITTLE KING."

hermit will return and" — He finished with a gasp; and at once fell to struggling so frantically with his bonds again, that he shook off the smothering sheepskin.

And now he heard the door open! The sound chilled him to the marrow — already he seemed to feel the knife at his throat. Horror made him close his eyes; horror made him open them again — and before him stood John Canty and Hugo!

He would have said "Thank God!" if his jaws had been free.

A moment or two later his limbs were at liberty, and his captors each gripping him by an arm, were hurrying him with all speed through the forest.

A VICTIM OF TREACHERY

CHAPTER XXII.

A VICTIM OF TREACHERY.

ONCE more "King Foo-Foo the First" was roving with the tramps and outlaws, a butt for their coarse jests and dull-witted railleries, and sometimes the victim of small spitefulnesses at the hands of Canty and Hugo when the Ruffler's back was turned. None but Canty and Hugo really disliked him. Some of the others liked him, and all admired his pluck and spirit. During two or three days, Hugo, in whose ward and charge the king was, did what he covertly could to make the boy uncomfortable; and at night, during the customary orgies, he amused the company by putting small indignities upon him — always as if by accident. Twice he stepped upon the king's toes — accidentally — and the king, as became his royalty, was contemptuously unconscious of it and indifferent to it; but the third time Hugo entertained himself in that way, the king felled him to the ground with a cudgel, to the prodigious delight of the tribe. Hugo, consumed with anger and shame, sprang up, seized a cudgel, and came at his small adversary in a fury. Instantly a ring was formed around the gladiators, and the betting and cheering began. But poor Hugo stood no chance whatever. His frantic and lubberly 'prentice-work found but a poor market for itself when pitted against an arm which had been trained by the first masters of Europe in single-stick, quarter-staff, and every art and trick of swordsmanship. The little king stood, alert but at graceful ease, and caught and turned aside the thick rain of blows with a facility and precision which set the motley on-lookers wild with admiration; and every now and then, when his practised eye detected an opening, and a lightning-swift rap upon Hugo's head followed as a result, the storm of cheers and laughter that swept the place was something wonderful to hear. At the end of fifteen minutes, Hugo, all battered, bruised, and the target for a pitiless bombardment of ridicule, slunk from the field; and the unscathed hero of the fight was seized and borne aloft upon the shoulders of the joyous rabble to the place of honor beside the Ruffler, where with vast ceremony he was crowned King of the Game-Cocks; his meaner title being at the same time solemnly cancelled and annulled, and a decree of banishment from the gang pronounced against any who should thenceforth utter it.

"HUGO STOOD NO CHANCE."

All attempts to make the king serviceable to the troop had failed. He had stubbornly refused to act; moreover he was always trying to escape. He had been thrust into an unwatched kitchen, the first day of his return; he not only came forth empty handed, but tried to rouse the housemates. He was sent out with a tinker to help him at his work; he would not work; moreover he threatened the tinker with his own soldering-iron; and finally both Hugo and the tinker found their hands full with the mere matter of keeping him from getting away. He delivered the thunders of his royalty upon the heads of all who hampered his liberties or tried to force him to service. He was sent out, in Hugo's charge, in company with a slatternly woman and a diseased baby, to beg; but the result was not encouraging — he declined to plead for the mendicants, or be a party to their cause in any way.

Thus several days went by; and the miseries of this tramping life, and the weariness and sordidness and meanness and vulgarity of it, became gradually and steadily so intolerable to the captive that he began at last to feel that his release from the hermit's knife must prove only a temporary respite from death, at best.

But at night, in his dreams, these things were forgotten, and he was on his throne, and master again. This, of course, intensified the sufferings of the awakening — so the mortifications of each succeeding morning of the few that passed between his return to bondage and the combat with Hugo, grew bitterer and bitterer, and harder and harder to bear.

The morning after that combat, Hugo got up with a heart filled with vengeful purposes against the king. He had two plans, in particular. One was to inflict upon the lad what would be, to his proud spirit and "imagined" royalty, a peculiar humiliation; and if he failed to accomplish this, his other plan was to put a crime of some kind upon the king and then betray him into the implacable clutches of the law.

In pursuance of the first plan, he purposed to put a "clime" upon the king's leg; rightly judging that that would mortify him to the last and perfect degree; and as soon as the clime should operate, he meant to get Canty's help, and *force* the king to expose his leg in the highway and beg for alms. "Clime" was the cant term for a sore, artificially created. To make a clime, the operator made a paste or poultice of unslaked lime, soap, and the rust of old iron, and spread it upon a piece of leather, which was then bound tightly upon the leg. This would presently fret off the skin, and make the flesh raw and

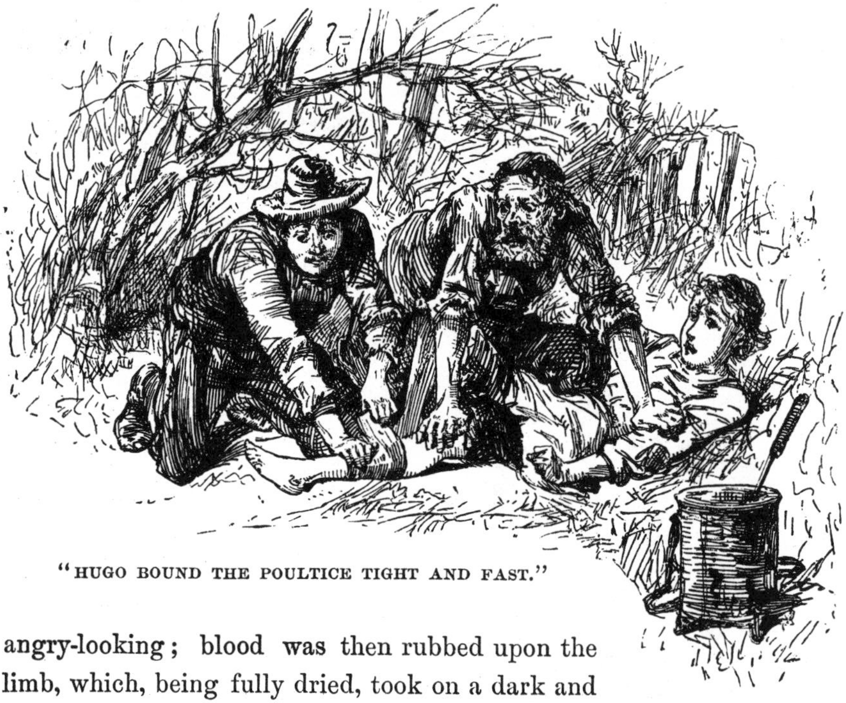

"HUGO BOUND THE POULTICE TIGHT AND FAST."

angry-looking; blood was then rubbed upon the limb, which, being fully dried, took on a dark and repulsive color. Then a bandage of soiled rags was put on in a cleverly careless way which would allow the hideous ulcer to be seen and move the compassion of the passer-by.[1]

Hugo got the help of the tinker whom the king had cowed with the soldering-iron; they took the boy out on a tinkering tramp, and as soon as they were out of sight of the camp they threw him down

[1] From "The English Rogue;" London, 1665.

and the tinker held him while Hugo bound the poultice tight and fast upon his leg.

The king raged and stormed, and promised to hang the two the moment the sceptre was in his hand again; but they kept a firm grip upon him and enjoyed his impotent struggling and jeered at his threats. This continued until the poultice began to bite; and in no long time its work would have been perfected, if there had been no interruption. But there was; for about this time the "slave" who had made the speech denouncing England's laws, appeared on the scene and put an end to the enterprise, and stripped off the poultice and bandage.

The king wanted to borrow his deliverer's cudgel and warm the jackets of the two rascals on the spot; but the man said no, it would bring trouble — leave the matter till night; the whole tribe being together, then, the outside world would not venture to interfere or interrupt. He marched the party back to camp and reported the affair to the Ruffler, who listened, pondered, and then decided that the king should not be again detailed to beg, since it was plain he was worthy of something higher and better — wherefore, on the spot he promoted him from the mendicant rank and appointed him to steal!

Hugo was overjoyed. He had already tried to make the king steal, and failed; but there would be no more trouble of that sort, now, for of course the king would not dream of defying a distinct command delivered directly from headquarters. So he planned a raid for that very afternoon, purposing to get the king in the law's grip in the course of it; and to do it, too, with such ingenious strategy, that it should seem to be accidental and unintentional; for the King of the Game-Cocks was popular, now, and the gang might not deal over-gently with an unpopular member who played so serious a treachery upon him as the delivering him over to the common enemy, the law.

Very well. All in good time Hugo strolled off to a neighboring village with his prey; and the two drifted slowly up and down one street after another, the one watching sharply for a sure chance to achieve his evil purpose, and the other watching as sharply for a chance to dart away and get free of his infamous captivity forever.

Both threw away some tolerably fair-looking opportunities; for both, in their secret hearts, were resolved to make absolutely sure work this time, and neither meant to allow his fevered desires to seduce him into any venture that had much uncertainty about it.

"TARRY HERE TILL I COME AGAIN."

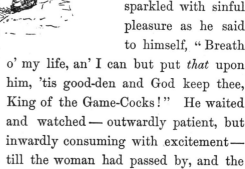

Hugo's chance came first. For at last a woman approached who carried a fat package of some sort in a basket. Hugo's eyes sparkled with sinful pleasure as he said to himself, "Breath o' my life, an' I can but put *that* upon him, 'tis good-den and God keep thee, King of the Game-Cocks!" He waited and watched — outwardly patient, but inwardly consuming with excitement — till the woman had passed by, and the time was ripe; then said, in a low voice —

"Tarry here till I come again," and darted stealthily after the prey.

The king's heart was filled with joy — he could make his escape, now, if Hugo's quest only carried him far enough away.

But he was to have no such luck. Hugo crept behind the woman, snatched the package, and came running back, wrapping it in an old piece of blanket which he carried on his arm. The hue and cry was raised in a moment, by the woman, who knew her loss by the lightening of her burden, although she had not seen the pilfering done. Hugo thrust the bundle into the king's hands without halting, saying, —

"Now speed ye after me with the rest, and cry 'Stop thief!' but mind ye lead them astray!"

The next moment Hugo turned a corner and darted down a crooked alley, — and in another moment or two he lounged into view again, looking innocent and indifferent, and took up a position behind a post to watch results.

The insulted king threw the bundle on the ground; and the blanket fell away from it just as the woman arrived, with an augmenting crowd at her heels; she seized the king's wrist with one hand, snatched up her bundle with the other, and began to pour out a tirade of abuse upon the boy while he struggled, without success, to free himself from her grip.

Hugo had seen enough — his enemy was captured and the law would get him, now — so he slipped away, jubilant and chuckling, and wended campwards, framing a judicious version of the matter to give to the Ruffler's crew as he strode along.

The king continued to struggle in the woman's strong grasp, and now and then cried out, in vexation —

"Unhand me, thou foolish creature; it was not I that bereaved thee of thy paltry goods."

The crowd closed around, threatening the king and calling him names; a brawny blacksmith in leather apron, and sleeves rolled to his elbows, made a reach for him, saying he would trounce him well, for a lesson; but just then a long sword flashed in the air and fell with convincing force upon the man's arm, flat-side down, the fantastic owner of it remarking pleasantly at the same time—

"Marry, good souls, let us proceed gently, not with ill blood and

"THE KING SPRANG TO HIS DELIVERER'S SIDE."

uncharitable words. This is matter for the law's consideration, not private and unofficial handling. Loose thy hold from the boy, goodwife."

The blacksmith averaged the stalwart soldier with a glance, then went muttering away, rubbing his arm; the woman released the boy's wrist reluctantly; the crowd eyed the stranger unlovingly, but prudently closed their mouths. The king sprang to his deliverer's side, with flushed cheeks and sparkling eyes, exclaiming —

"Thou hast lagged sorely, but thou comest in good season, now, Sir Miles; carve me this rabble to rags!"

THE PRINCE
A PRISONER

CHAPTER XXIII.

THE PRINCE A PRISONER.

HENDON forced back a smile, and bent down and whispered in the king's ear —

"Softly, softly, my prince, wag thy tongue warily — nay, suffer it not to wag at all. Trust in me — all shall go well in the end." Then he added, to himself: "*Sir* Miles! Bless me, I had totally forgot I was a knight! Lord how marvellous a thing it is, the grip his memory doth take upon his quaint and crazy fancies! . . . An empty and foolish title is mine, and yet it is something to have deserved it, for I think it is more honor to be held worthy to be a spectre-knight in his Kingdom of Dreams and Shadows, than to be held base enough to be an earl in some of the *real* kingdoms of this world."

The crowd fell apart to admit a constable, who approached and was about to lay his hand upon the king's shoulder, when Hendon said —

"Gently, good friend, withhold your hand — he shall go peaceably; I am responsible for that. Lead on, we will follow."

The officer led, with the woman and her bundle; Miles and the king followed after, with the crowd at their heels. The King was inclined to rebel; but Hendon said to him in a low voice —

"Reflect, sire — your laws are the wholesome breath of your own royalty; shall their source resist them, yet require the branches to respect them? Apparently one of these laws has been broken; when the king is on his throne again, can it ever grieve him to remember that when he was seemingly a private person he loyally sunk the king in the citizen and submitted to its authority?"

"Thou art right; say no more; thou shalt see that whatsoever the king of England requires a subject to suffer under the law, he will himself suffer while he holdeth the station of a subject."

"GENTLY, GOOD FRIEND."

When the woman was called upon to testify before the justice of the peace, she swore that the small prisoner at the bar was the person who had committed the theft; there was none able to show the contrary, so the king stood convicted. The bundle was now unrolled, and when the contents proved to be a plump little dressed pig, the judge looked troubled, whilst Hendon turned pale, and his body was thrilled with an electric shiver of dismay; but the king remained unmoved, protected by his ignorance. The judge meditated, during an ominous pause, then turned to the woman, with the question —

"What dost thou hold this property to be worth?"

The woman courtesied and replied —

"Three shillings and eightpence, your worship — I could not abate a penny and set forth the value honestly."

The justice glanced around uncomfortably upon the crowd, then nodded to the constable and said —

"Clear the court and close the doors."

It was done. None remained but the two officials, the accused, the accuser, and Miles Hendon. This latter was rigid and colorless, and on his forehead big drops of cold sweat gathered, broke and blended together, and trickled down his face. The judge turned to the woman again, and said, in a compassionate voice —

"'Tis a poor ignorant lad, and mayhap was driven hard by hunger, for these be grievous times for the unfortunate; mark you, he hath not an evil face — but when hunger driveth — Good woman! dost know that when one steals a thing above the value of thirteen pence ha'penny the law saith he shall *hang* for it!"

The little king started, wide-eyed with consternation, but controlled himself and held his peace; but not so the woman. She sprang to her feet, shaking with fright, and cried out —

"O, good lack, what have I done! God-a-mercy, I would not hang the poor thing for the whole world! Ah, save me from this, your worship — what shall I do, what *can* I do?"

The justice maintained his judicial composure, and simply said —

"Doubtless it is allowable to revise the value, since it is not yet writ upon the record."

"Then in God's name call the pig eightpence, and heaven bless the day that freed my conscience of this awesome thing!"

Miles Hendon forgot all decorum in his delight; and surprised the king and wounded his dignity, by throwing his arms around him and hugging him. The woman made her grateful adieux and started away with her pig; and when the constable opened the door for her, he followed her out into the narrow hall. The justice proceeded to write in his record book. Hendon, always alert, thought he would like to know why the officer followed the woman out; so he slipped softly

"SHE SPRANG TO HER FEET."

into the dusky hall and listened. He heard a conversation to this effect —

"It is a fat pig, and promises good eating; I will buy it of thee; here is the eightpence."

"Eightpence, indeed! Thou'lt do no such thing. It cost me three shillings and eightpence, good honest coin of the last reign, that old Harry that's just dead ne'er touched nor tampered with. A fig for thy eightpence!"

"Stands the wind in that quarter? Thou wast under oath, and so swore falsely when thou saidst the value was but eightpence. Come straightway back with me before his worship, and answer for the crime!—and then the lad will hang."

"There, there, dear heart, say no more, I am content. Give me the eightpence, and hold thy peace about the matter."

The woman went off crying; Hendon slipped back into the court room, and the constable presently followed, after hiding his prize in some convenient place. The justice wrote a while longer, then read the king a wise and kindly lecture, and sentenced him to a short imprisonment in the common jail, to be followed by a public flogging. The astounded king opened his mouth and was probably going to order the good judge to be beheaded on the spot; but he caught a warning sign from Hendon, and succeeded in closing his mouth again before he lost any thing out of it. Hendon took him by the hand, now, made reverence to the justice, and the two departed in the wake of the constable toward the jail. The moment the street was reached, the inflamed monarch halted, snatched away his hand, and exclaimed—

"Idiot, dost imagine I will enter a common jail *alive?*"

Hendon bent down and said, somewhat sharply—

"*Will* you trust in me? Peace! and forbear to worsen our chances with dangerous speech. What God wills, will happen; thou canst not hurry it, thou canst not alter it; therefore wait, and be patient—'twill be time enow to rail or rejoice when what is to happen has happened."[1]

[1] See Notes to Chapter 23, at end of volume.

The Escape

CHAPTER XXIV.

THE ESCAPE.

THE short winter day was nearly ended. The streets were deserted, save for a few random stragglers, and these hurried straight along, with the intent look of people who were only anxious to accomplish their errands as quickly as possible and then snugly house themselves from the rising wind and the gathering twilight. They looked neither to the right nor to the left; they paid no attention to our party, they did not even seem to see them. Edward the Sixth wondered if the spectacle of a king on his way to jail had ever encountered such marvellous indifference before. By and by the constable arrived at a deserted market-square and proceeded to cross it. When he had reached the middle of it, Hendon laid his hand upon his arm, and said in a low voice —

"Bide a moment, good sir, there is none in hearing, and I would say a word to thee."

"My duty forbids it, sir; prithee hinder me not, the night comes on."

"Stay, nevertheless, for the matter concerns thee nearly. Turn thy back a moment and seem not to see: *let this poor lad escape.*"

"This to me, sir! I arrest thee in" —

"Nay, be not too hasty. See thou be careful and commit no foolish error" — then he shut his voice down to a whisper, and said in the man's ear — "the pig thou hast purchased for eightpence may cost thee thy neck, man!"

The poor constable, taken by surprise, was speechless, at first, then found his tongue and fell to blustering and threatening; but Hendon was tranquil, and waited with patience till his breath was spent; then said —

"I have a liking to thee, friend, and would not willingly see thee come to harm. Observe, I heard it all — every word. I will prove it to thee." Then he repeated the conversation which the officer and the woman had had together in the hall, word for word, and ended with —

"THE PIG MAY COST THY NECK, MAN."

"There — have I set it forth correctly? Should not I be able to set it forth correctly before the judge, if occasion required?"

The man was dumb with fear and distress, for a moment; then he rallied and said with forced lightness —

"'Tis making a mighty matter indeed, out of a jest; I but plagued the woman for mine amusement."

"Kept you the woman's pig for amusement?"

The man answered sharply —

"Nought else, good sir — I tell thee 'twas but a jest."

"I do begin to believe thee," said Hendon, with a perplexing mixture of mockery and half-conviction in his tone; "but tarry thou here a moment whilst I run and ask his worship — for nathless, he being a man experienced in law, in jests, in" —

He was moving away, still talking; the constable hesitated, fidgetted, spat out an oath or two, then cried out —

"Hold, hold, good sir — prithee wait a little — the judge! why man, he hath no more sympathy with a jest than hath a dead corpse! — come, and we will speak further. Ods body! I seem to be in evil case — and all for an innocent and thoughtless pleasantry. I am a man of family; and my wife and little ones — List to reason, good your worship: what wouldst thou of me?"

"Only that thou be blind and dumb and paralytic whilst one may count a hundred thousand — counting slowly," said Hendon, with the expression of a man who asks but a reasonable favor, and that a very little one.

"It is my destruction!" said the constable despairingly. "Ah, be reasonable, good sir; only look at this matter, on all its sides, and see how mere a jest it is — how manifestly and how plainly it is so. And even if one granted it were not a jest, it is a fault so small that e'en the grimmest penalty it could call forth would be but a rebuke and warning from the judge's lips."

Hendon replied with a solemnity which chilled the air about him —

"This jest of thine hath a name, in law, — wot you what it is?"

"I knew it not! Peradventure I have been unwise. I never dreamed it had a name — ah, sweet heaven, I thought it was original."

"Yes, it hath a name. In the law this crime is called *Non compos mentis lex talionis sic transit gloria Mundi.*"

"Ah, my God!"

"And the penalty is death!"

"God be merciful to me, a sinner!"

"By advantage taken of one in fault, in dire peril, and at thy mercy, thou hast seized goods worth above thirteen pence ha'penny, paying but a trifle for the same; and this, in the eye of the law, is constructive barratry, misprision of treason, malfeasance in office, *ad hominem expurgatis in statu quo* — and the penalty is death by the halter, without ransom, commutation, or benefit of clergy."

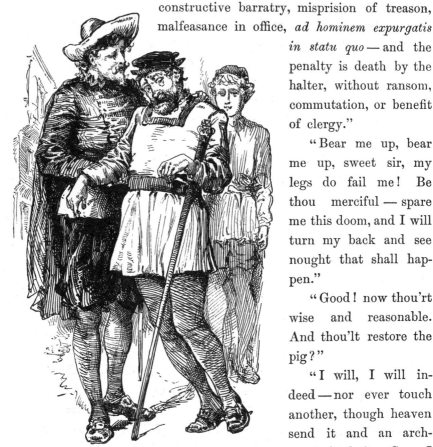

"BEAR ME UP, BEAR ME UP, SWEET SIR!"

"Bear me up, bear me up, sweet sir, my legs do fail me! Be thou merciful — spare me this doom, and I will turn my back and see nought that shall happen."

"Good! now thou'rt wise and reasonable. And thou'lt restore the pig?"

"I will, I will indeed — nor ever touch another, though heaven send it and an archangel fetch it. Go — I am blind for thy sake — I see nothing. I will say thou didst break in and wrest the prisoner from my hands by force. It is but a crazy, ancient door — I will batter it down myself betwixt midnight and the morning."

"Do it, good soul, no harm will come of it; the judge hath a loving charity for this poor lad, and will shed no tears and break no jailer's bones for his escape."

HENDON HALL

CHAPTER XXV.

HENDON HALL.

As soon as Hendon and the king were out of sight of the constable, his majesty was instructed to hurry to a certain place outside the town, and wait there, whilst Hendon should go to the inn and settle his account. Half an hour later the two friends were blithely jogging eastward on Hendon's sorry steeds. The king was warm and comfortable, now, for he had cast his rags and clothed himself in the second-hand suit which Hendon had bought on London Bridge.

Hendon wished to guard against over-fatiguing the boy; he judged that hard journeys, irregular meals, and illiberal measures of sleep would be bad for his crazed mind; whilst rest, regularity, and moderate exercise would be pretty sure to hasten its cure; he longed to see the stricken intellect made well again and its diseased visions driven out of the tormented little head; therefore he resolved to move by easy stages toward the home whence he had so long been banished, instead of obeying the impulse of his impatience and hurrying along night and day.

When he and the king had journeyed about ten miles, they reached a considerable village, and halted there for the night, at a good inn. The former relations were resumed; Hendon stood behind the king's chair, while he dined, and waited upon him; undressed him when he was ready for bed; then took the floor for his own quarters, and slept athwart the door, rolled up in a blanket.

The next day, and the day after, they jogged lazily along talking over the adventures they had met since their separation, and mightily enjoying each other's narratives. Hendon detailed all his wide wanderings in search of the king, and described how the archangel had led him a fool's journey all over the forest, and taken him back to the hut, finally, when he found he could not get rid of him. Then — he said

"JOGGING EASTWARD ON SORRY STEEDS."

— the old man went into the bedchamber and came staggering back looking broken-hearted, and saying he had expected to find that the boy had returned and lain down in there to rest, but it was not so. Hendon had waited at the hut all day; hope of the king's return died out, then, and he departed upon the quest again.

"And old Sanctum Sanctorum *was* truly sorry your highness came not back," said Hendon; "I saw it in his face."

"Marry I will never doubt *that!*" said the King — and then told his own story; after which, Hendon was sorry he had not destroyed the archangel.

During the last day of the trip, Hendon's spirits were soaring. His tongue ran constantly. He talked about his old father, and his brother Arthur, and told of many things which illustrated their high and generous characters; he went

"THERE IS THE VILLAGE, MY PRINCE!"

into loving frenzies over his Edith, and was so gladhearted that he was even able to say some gentle and brotherly things about Hugh. He dwelt a deal on the coming meeting at Hendon Hall; what a surprise it would be to everybody, and what an outburst of thanksgiving and delight there would be.

It was a fair region, dotted with cottages and orchards, and the road led through broad pasture lands whose receding expanses, marked with gentle elevations and depressions, suggested the swelling and subsiding undulations of the sea. In the afternoon the returning prodigal made constant deflections from his course to see if by ascending some hillock he might not pierce the distance and catch a glimpse of his home. At last he was successful, and cried out excitedly —

"There is the village, my prince, and there is the Hall close by! You may see the towers from here; and that wood there — that is my father's park. Ah, *now* thou'lt know what state and grandeur be! A house with seventy rooms — think of that! — and seven and twenty servants! A brave lodging for such as we, is it not so? Come, let us speed — my impatience will not brook further delay."

All possible hurry was made; still, it was after three o'clock before the village was reached. The travellers scampered through it, Hendon's tongue going all the time. "Here is the church — covered with the same ivy — none gone, none added." "Yonder is the inn, the old Red Lion, — and yonder is the marketplace." "Here is the Maypole, and here the pump — nothing is altered; nothing but the people, at any rate; ten years make a change in people; some of these I seem to know, but none know me." So his chat ran on. The end of the village was soon reached; then the travellers struck into a crooked, narrow road, walled in with tall hedges, and hurried briskly along it for a half mile, then passed into a vast flower garden through an imposing gateway whose huge stone pillars bore sculptured armorial devices. A noble mansion was before them.

"Welcome to Hendon Hall, my king!" exclaimed Miles. "Ah, 'tis a great day! My father and my brother, and the lady Edith will be so mad with joy that they will have eyes and tongue for none but me in the first transports of the meeting, and so thou'lt seem but coldly welcomed — but mind it not; 'twill soon seem otherwise; for when I say thou art my ward, and tell them how costly is my love for thee, thou'lt see them take thee to their breasts for Miles Hendon's sake, and make their house and hearts thy home forever after!"

The next moment Hendon sprang to the ground before the great door, helped the king down, then took him by the hand and rushed within. A few steps brought him to a spacious apartment; he entered, seated the king with more hurry than

"'EMBRACE ME, HUGH,' HE CRIED."

ceremony, then ran toward a young man who sat at a writing table in front of a generous fire of logs.

"Embrace me, Hugh," he cried, "and say thou'rt glad I am come again! and call our father, for home is not home till I shall touch his hand, and see his face, and hear his voice once more!"

But Hugh only drew back, after betraying a momentary surprise, and bent a grave stare upon the intruder — a stare which indicated somewhat of offended dignity, at first, then changed, in response to some inward thought or purpose, to an expression of marvelling curiosity, mixed with a real or assumed compassion. Presently he said, in a mild voice —

"Thy wits seem touched, poor stranger; doubtless thou hast suffered privations and rude buffetings at the world's hands; thy looks and dress betoken it. Whom dost thou take me to be?"

"Take thee? Prithee for whom else than whom thou art? I take thee to be Hugh Hendon," said Miles, sharply.

The other continued, in the same soft tone —

"And whom dost thou imagine thyself to be?"

"Imagination hath nought to do with it! Dost thou pretend thou knowest me not for thy brother Miles Hendon?"

An expression of pleased surprise flitted across Hugh's face, and he exclaimed —

"What! thou art not jesting? can the dead come to life? God be praised if it be so! Our poor lost boy restored to our arms after all these cruel years! Ah, it seems too good to be true, it *is* too good to be true — I charge thee, have pity, do not trifle with me! Quick — come to the light — let me scan thee well!"

He seized Miles by the arm, dragged him to the window, and began to devour him from head to foot with his eyes, turning him this way and that, and stepping briskly around him and about him to prove him from all points of view; whilst the returned prodigal, all aglow with gladness, smiled, laughed, and kept nodding his head and saying —

"Go on, brother, go on, and fear not; thou'lt find nor limb nor feature that cannot bide the test. Scour and scan me to thy content, my good old Hugh — I am indeed thy old Miles, thy same old Miles, thy lost brother, is't not so? Ah, 'tis a great day — I *said* 'twas a great day! Give me thy hand, give me thy cheek — lord, I am like to die of very joy!"

He was about to throw himself upon his brother; but Hugh put up his hand in dissent, then dropped his chin mournfully upon his breast, saying with emotion —

"Ah, God of his mercy give me strength to bear this grievous disappointment!"

Miles, amazed, could not speak, for a moment; then he found his tongue, and cried out —

"HUGH PUT UP HIS HAND IN DISSENT."

"*What* disappointment? Am I not thy brother?"

Hugh shook his head sadly, and said —

"I pray heaven it may prove so, and that other eyes may find the resemblances that are hid from mine. Alack, I fear me the letter spoke but too truly."

"What letter?"

"One that came from over sea, some six or seven years ago. It said my brother died in battle."

"It was a lie! Call thy father — he will know me."

"One may not call the dead."

"Dead?" Miles's voice was subdued, and his lips trembled. "My father dead! — O, this is heavy news. Half my new joy is withered now. Prithee let me see my brother Arthur — he will know me; he will know me and console me."

"He, also, is dead."

"God be merciful to me, a stricken man! Gone, — both gone — the worthy taken and the worthless spared, in me! Ah! I crave your mercy! — do not say the lady Edith" —

"Is dead? No, she lives."

"Then, God be praised, my joy is whole again! Speed thee, brother — let her come to me! An' *she* say I am not myself, — but she will not; no, no, *she* will know me, I were a fool to doubt it. Bring her — bring the old servants; they, too, will know me."

"All are gone but five — Peter, Halsey, David, Bernard and Margaret."

So saying, Hugh left the room. Miles stood musing, a while, then began to walk the floor, muttering —

"The five arch villains have survived the two-and-twenty leal and honest — 'tis an odd thing."

He continued walking back and forth, muttering to himself; he had forgotten the king entirely. By and by his majesty said gravely, and with a touch of genuine compassion, though the words themselves were capable of being interpreted ironically —

"Mind not thy mischance, good man; there be others in the world whose identity is denied, and whose claims are derided. Thou hast company."

"Ah, my king," cried Hendon, coloring slightly, "do not thou condemn me — wait, and thou shalt see. I am no impostor — she will say it; you shall hear it from the sweetest lips in England. I an impostor? Why I know this old hall, these pictures of my ancestors, and all these things that are about us, as a child knoweth its own nursery. Here was I born and bred, my lord; I speak the truth; I would not deceive thee; and should none else believe, I pray thee do not *thou* doubt me — I could not bear it."

"A BEAUTIFUL LADY, RICHLY CLOTHED, FOLLOWED HUGH."

"I do not doubt thee," said the king, with a childlike simplicity and faith.

"I thank thee out of my heart!" exclaimed Hendon, with a fervency which showed that he was touched. The king added, with the same gentle simplicity —

"Dost thou doubt *me?*"

A guilty confusion seized upon Hendon, and he was grateful that the door opened to admit Hugh, at that moment, and saved him the necessity of replying.

A beautiful lady, richly clothed, followed Hugh, and after her came several liveried servants. The lady walked slowly, with her head bowed and her eyes fixed upon the floor. The face was unspeakably sad. Miles Hendon sprang forward, crying out —

"O, my Edith, my darling" —

But Hugh waved him back, gravely, and said to the lady —

"Look upon him. Do you know him?"

At the sound of Miles's voice the woman had started, slightly, and her cheeks had flushed; she was trembling, now. She stood still, during an impressive pause of several moments; then slowly lifted up her head and looked into Hendon's eyes with a stony and frightened gaze; the blood sank out of her face, drop by drop, till nothing remained but the gray pallor of death; then she said, in a voice as dead as the face, "I know him not!" and turned, with a moan and a stifled sob, and tottered out of the room.

Miles Hendon sank into a chair and covered his face with his hands. After a pause, his brother said to the servants —

"You have observed him. Do you know him?"

They shook their heads; then the master said —

"The servants know you not, sir. I fear there is some mistake. You have seen that my wife knew you not."

"Thy *wife!*" In an instant Hugh was pinned to the wall, with an iron grip about his throat. "O, thou fox-hearted slave, I see it all! Thou'st writ the lying letter thyself, and my stolen bride and goods are its fruit. There — now get thee gone, lest I shame mine honorable soldiership with the slaying of so pitiful a manikin!"

Hugh, red-faced, and almost suffocated, reeled to the nearest chair, and commanded the servants to seize and bind the murderous stranger. They hesitated, and one of them said —

"He is armed, Sir Hugh, and we are weaponless."

"Armed? What of it, and ye so many? Upon him, I say!"

But Miles warned them to be careful what they did, and added —

"Ye know me of old — I have not changed; come on, an' it like you."

This reminder did not hearten the servants much; they still held back.

"Then go, ye paltry cowards, and arm yourselves and guard the doors, whilst I send one to fetch the watch;" said Hugh. He turned, at the threshold, and said to Miles,

"HUGH WAS PINNED TO THE WALL."

"You'll find it to your advantage to offend not with useless endeavors at escape."

"Escape? Spare thyself discomfort, an' that is all that troubles thee. For Miles Hendon is master of Hendon Hall and all its belongings. He will remain —doubt it not."

DISOWNED

CHAPTER XXVI.

DISOWNED.

THE king sat musing a few moments, then looked up and said—

"'Tis strange—most strange. I cannot account for it."

"No, it is not strange, my liege. I know him, and this conduct is but natural. He was a rascal from his birth."

"O, I spake not of *him*, Sir Miles."

"Not of him? Then of what? What is it that is strange?"

"That the king is not missed."

"How? Which? I doubt I do not understand."

"Indeed? Doth it not strike you as being passing strange that the land is not filled with couriers and proclamations describing my person and making search for me? Is it no matter for commotion and distress that the head of the State is gone?—that I am vanished away and lost?"

"Most true, my king, I had forgot." Then Hendon sighed, and muttered to himself, "Poor ruined mind—still busy with its pathetic dream."

"But I have a plan that shall right us both. I will write a paper, in three tongues—Latin, Greek and English—and thou shalt haste away with it to London in the morning. Give it to none but my uncle, the lord Hertford; when he shall see it, he will know and say I wrote it. Then he will send for me."

"Might it not be best, my prince, that we wait, here, until I prove myself and make my rights secure to my domains? I should be so much the better able then to"—

The king interrupted him imperiously—

"Peace! What are thy paltry domains, thy trivial interests, contrasted with matters which concern the weal of a nation and the integrity of a throne!" Then he added, in a gentle voice, as if he were sorry

"OBEY, AND HAVE NO FEAR."

for his severity, "Obey, and have no fear; I will right thee, I will make thee whole—yes, more than whole. I shall remember, and requite."

So saying, he took the pen, and set himself to work. Hendon contemplated him lovingly, a while, then said to himself—

"An' it were dark, I should think it *was* a king that spoke; there's no denying it, when the humor's upon him he doth thunder and lighten like your true king — now where got he that trick? See him scribble and scratch away contentedly at his meaningless pot-hooks, fancying them to be Latin and Greek — and except my wit shall serve me with a lucky device for diverting him from his purpose, I shall be forced to pretend to post away to-morrow on this wild errand he hath invented for me."

The next moment Sir Miles's thoughts had gone back to the recent episode. So absorbed was he in his musings, that when the king presently handed him the paper which he had been writing, he received it and pocketed it without being conscious of the act. "How marvellous strange she acted," he muttered. "I think she knew me — and I think she did *not* know me. These opinions do conflict, I perceive it plainly; I cannot reconcile them, neither can I, by argument, dismiss either of the two, or even persuade one to outweigh the other. The matter standeth simply thus: she *must* have known my face, my figure, my voice, for how could it be otherwise? yet she *said* she knew me not, and that is proof perfect, for she cannot lie. But stop — I think I begin to see. Peradventure he hath influenced her — commanded her — compelled her, to lie. That is the solution! The riddle is unriddled. She seemed dead with fear — yes, she was under his compulsion. I will seek her; I will find her; now that he is away, she will speak her true mind. She will remember the old times when we were little playfellows together, and this will soften her heart, and she will no more betray me, but will confess me. There is no treacherous blood in her — no, she was always honest and true. She has loved me, in those old days — this is my security; for whom one has loved, one cannot betray."

He stepped eagerly toward the door; at that moment it opened, and the lady Edith entered. She was very pale, but she walked with a firm step, and her carriage was full of grace and gentle dignity. Her face was as sad as before.

Miles sprang forward, with a happy confidence, to meet her, but she checked him with a hardly perceptible gesture, and he stopped where he was. She seated herself, and asked him to do likewise. Thus simply did she take the sense of old-comradeship out of him, and transform him into a stranger and a guest. The surprise of it, the bewildering unexpectedness of it, made him begin to question, for a moment, if he *was* the person he was pretending to be, after all. The lady Edith said—

"Sir, I have come to warn you. The mad cannot be persuaded out of their delusions, perchance; but doubtless they may be persuaded to avoid perils. I think this dream of yours hath the seeming of honest truth to you, and therefore is not criminal—but do not tarry here with it; for here it is dangerous." She looked steadily into Miles's face, a moment, then added, impressively, "It is the more dangerous for that you *are* much like what our lost lad must have grown to be, if he had lived."

"Heavens, madam, but I *am* he!"

"I truly think you think it, sir. I question not your honesty in that—I but warn you, that is all. My husband is master in this region; his power hath hardly any limit; the people prosper or starve, as he wills. If you resembled not the man whom you profess to be, my husband might bid you pleasure yourself with your dream in peace; but trust me, I know him well, I know what he will do; he will say to all, that you are but a mad impostor, and straightway all will echo him." She bent upon Miles that same steady look once more, and added: "If you *were* Miles Hendon, and he knew it and all the region knew it—consider what I am saying, weigh it well—you would stand in the same peril, your punishment would be no less sure; he would deny you and denounce you, and none would be bold enough to give you countenance."

"Most truly I believe it," said Miles, bitterly. "The power that can command one life-long friend to betray and disown another, and be obeyed, may well look to be obeyed in quarters where bread and life are on the stake and no cobweb ties of loyalty and honor are concerned."

A faint tinge appeared for a moment in the lady's cheek, and she dropped her eyes to the floor; but her voice be-

"AM I MILES HENDON?"

trayed no emotion when she proceeded—

"I have warned you, I must still warn you, to go hence. This man will destroy you, else. He is a tyrant who knows no pity. I, who am his fettered slave, know this. Poor Miles, and Arthur, and my dear guardian, Sir Richard, are free of him, and at rest—better that you were with them than that you bide here in the clutches of this miscreant. Your pretensions are a menace to his title and possessions; you have assaulted him in his own house — you are ruined if you stay. Go — do not hesitate. If you lack money, take this purse, I beg of you, and bribe the servants to let you pass. O be warned, poor soul, and escape while you may."

Miles declined the purse with a gesture, and rose up and stood before her.

"Grant me one thing," he said. "Let your eyes rest upon mine, so that I may see if they be steady. There — now answer me. Am I Miles Hendon?"

"No. I know you not."

"Swear it!"

The answer was low, but distinct —

"I swear."

"O, this passes belief!"

"Fly! Why will you waste the precious time? Fly, and save yourself."

At that moment the officers burst into the room and a violent struggle began; but Hendon was soon overpowered and dragged away. The king was taken, also, and both were bound, and led to prison.

IN PRISON

CHAPTER XXVII.

IN PRISON.

THE cells were all crowded; so the two friends were chained in a large room where persons charged with trifling offences were commonly kept. They had company, for there were some twenty manacled and fettered prisoners here, of both sexes and of varying ages, — an obscene and noisy gang. The king chafed bitterly over the stupendous indignity thus put upon his royalty, but Hendon was moody and taciturn. He was pretty thoroughly bewildered. He had come home, a jubilant prodigal, expecting to find everybody wild with joy over his return; and instead had got the cold shoulder and a jail. The promise and the fulfilment differed so widely, that the effect was stunning; he could not decide whether it was most tragic or most grotesque. He felt much as a man might who had danced blithely out to enjoy a rainbow, and got struck by lightning.

But gradually his confused and tormenting thoughts settled down into some sort of order, and then his mind centred itself upon Edith. He turned her conduct over, and examined it in all lights, but he could not make any thing satisfactory out of it. Did she know him? — or didn't she know him? It was a perplexing puzzle, and occupied him a long time; but he ended, finally, with the conviction that she did know him, and had repudiated him for interested reasons. He wanted to load her name with curses now; but this name had so long been sacred to him that he found he could not bring his tongue to profane it.

Wrapped in prison blankets of a soiled and tattered condition, Hendon and the king passed a troubled night. For a bribe the jailer had furnished liquor to some of the prisoners; singing of ribald songs, fighting, shouting, and carousing, was the natural consequence. At last, a while after midnight, a man attacked a woman and nearly killed her by beating her over the head with his manacles before the jailer could come to the rescue. The jailer restored peace by giving the man a sound clubbing about the head and shoulders—then the carousing ceased; and after that, all had an opportunity to sleep who did not mind the annoyance of the moanings and groanings of the two wounded people.

"CHAINED IN A LARGE ROOM."

During the ensuing week, the days and nights were of a monotonous sameness, as to events; men whose faces Hendon remembered more or less distinctly, came, by day, to gaze at the "impostor" and repudiate and insult him; and by night the carousing and brawling went on, with symmetrical regularity. However, there was a change of incident at last. The jailer brought in an old man, and said to him—

"The villain is in this room—cast thy old eyes about and see if thou canst say which is he."

Hendon glanced up, and experienced a pleasant sensation for the first time since he had been in the jail. He said to himself, "This is Blake Andrews, a servant all his life in my father's family — a good honest soul, with a right heart in his breast. That is, formerly. But none are true, now; all are liars. This man will know me — and will deny me, too, like the rest."

The old man gazed around the room, glanced at each face in turn, and finally said —

"I see none here but paltry knaves, scum o' the streets. Which is he?"

The jailer laughed.

"Here," he said; "scan this big animal, and grant me an opinion."

The old man approached, and looked Hendon over, long and earnestly, then shook his head and said —

"Marry, *this* is no Hendon — nor ever was!"

"Right! Thy old eyes are sound yet. An' I were Sir Hugh, I would take the shabby carle and" —

The jailer finished by lifting himself a-tip-toe with an imaginary halter, at the same time making a gurgling noise in his throat suggestive of suffocation. The old man said, vindictively —

"Let him bless God an' he fare no worse. An' *I* had the handling o' the villain, he should roast, or I am no true man!"

The jailer laughed a pleasant hyena laugh, and said —

"Give him a piece of thy mind, old man — they all do it. Thou'lt find it good diversion."

Then he sauntered toward his ante-room and disappeared. The old man dropped upon his knees and whispered —

"God be thanked, thou'rt come again, my master! I believed thou wert dead these seven years, and lo, here thou art alive! I knew thee the moment I saw thee; and main hard work it was to keep a stony countenance and seem to see none here but tuppenny knaves and rubbish o' the streets. I am old and poor, Sir Miles; but say the word and I will go forth and proclaim the truth though I be strangled for it."

"No," said Hendon; "thou shalt not. It would ruin thee, and yet

"THE OLD MAN LOOKED HENDON OVER."

help but little in my cause. But I thank thee; for thou hast given me back somewhat of my lost faith in my kind."

The old servant became very valuable to Hendon and the king; for he dropped in several times a day to "abuse" the former, and always smuggled in a few delicacies to help out the prison bill of fare; he also furnished the current news. Hendon reserved the dainties for the king; without them his majesty might not have survived, for he was not able to eat the coarse and wretched food provided by the jailer. Andrews was obliged to confine himself to brief visits, in order to avoid suspicion; but he managed to impart a fair degree of information each time — information delivered in a low voice, for Hendon's benefit, and interlarded with insulting epithets delivered in a louder voice, for the benefit of other hearers.

"INFORMATION DELIVERED IN A LOW VOICE."

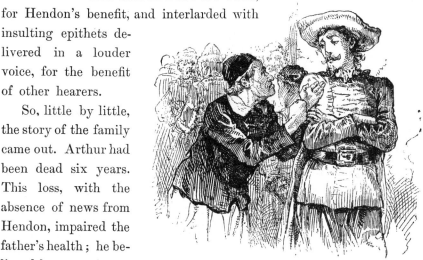

So, little by little, the story of the family came out. Arthur had been dead six years. This loss, with the absence of news from Hendon, impaired the father's health; he believed he was going to die, and he wished to see Hugh and Edith settled in life before he passed away; but Edith begged hard for delay, hoping for Miles's return; then the letter came which brought the news of Miles's death; the shock prostrated Sir Richard; he believed his end was very near, and he and Hugh insisted upon the marriage; Edith begged for and obtained a month's respite; then another, and finally a third; the marriage then took place, by the death-bed of Sir Richard. It had not proved a happy one. It was whispered about the country that shortly after the nuptials the bride found among her husband's papers several rough and incomplete drafts of the fatal letter, and had accused him of precipitating the marriage — and Sir Richard's death, too — by a wicked forgery. Tales of cruelty to the lady Edith and the servants were to be heard on all hands; and since the father's death Sir Hugh had thrown off all soft disguises and become a pitiless master toward all who in any way depended upon him and his domains for bread.

There was a bit of Andrews's gossip which the king listened to with a lively interest —

"There is rumor that the king is mad. But in charity forbear to say *I* mentioned it, for 'tis death to speak of it, they say."

His majesty glared at the old man and said —

"The king is *not* mad, good man — and thou'lt find it to thy advantage to busy thyself with matters that nearer concern thee than this seditious prattle."

"What doth the lad mean?" said Andrews, surprised at this brisk assault from such an unexpected quarter. Hendon gave him a sign, and he did not pursue his question, but went on with his budget —

"The late king is to be buried at Windsor in a day or two — the 16th of the month, — and the new king will be crowned at Westminster the 20th."

"Methinks they must needs find him first," muttered his majesty; then added, confidently, "but they will look to that — and so also shall I."

"In the name of" —

But the old man got no further — a warning sign from Hendon checked his remark. He resumed the thread of his gossip —

"Sir Hugh goeth to the coronation — and with grand hopes. He confidently looketh to come back a peer, for he is high in favor with the Lord Protector."

"What Lord Protector?" asked his majesty.

"His grace the Duke of Somerset."

"What Duke of Somerset?"

"Marry, there is but one—Seymour, earl of Hertford."

The king asked, sharply—

"Since when is *he* a duke, and Lord Protector?"

"Since the last day of January."

"And prithee who made him so?"

"Himself and the Great Council—with help of the king."

His majesty started violently. "The *king!*" he cried. "*What* king, good sir?"

"What king, indeed! (God-a-mercy, what aileth the boy?) Sith we have but one, 'tis not difficult to answer—his most sacred majesty King Edward the Sixth—whom God preserve! Yea, and a dear and gracious little urchin is he, too; and whether he be mad or no—and they say he mendeth daily—his praises are on all men's lips; and all bless him, likewise, and offer prayers that he may be spared to reign long in England; for he began humanely, with saving the old duke of Norfolk's life, and now is he bent on destroying the cruelest of the laws that harry and oppress the people."

"THE KING!" HE CRIED. "WHAT KING?"

This news struck his majesty dumb with amazement, and plunged him into so deep and dismal a revery that he heard no more of the old man's gossip. He wondered if the "little urchin" was the beggar-boy whom he left dressed in his own garments in the palace. It did not seem possible that this could be, for surely his manners and speech would betray him if he pretended to be the prince of Wales — then he would be driven out, and search made for the true prince. Could it be that the Court had set up some sprig of the nobility in his place? No, for his uncle would not allow that — he was all-powerful and could and would crush such a movement, of course. The boy's musings profited him nothing; the more he tried to unriddle the mystery the more perplexed he became, the more his head ached, and the worse he slept. His impatience to get to London grew hourly, and his captivity became almost unendurable.

Hendon's arts all failed with the king — he could not be comforted, but a couple of women who were chained near him, succeeded better. Under their gentle ministrations he found peace and learned a degree of patience. He was very grateful, and came to love them dearly and to delight in the sweet and soothing influence of their presence. He asked them why they were in prison, and when they said they were Baptists, he smiled, and inquired —

"Is that a crime to be shut up for, in a prison? Now I grieve, for I shall lose ye — they will not keep ye long for such a little thing."

They did not answer; and something in their faces made him uneasy. He said, eagerly —

"You do not speak — be good to me, and tell me — there will be no other punishment? Prithee tell me there is no fear of that."

They tried to change the topic, but his fears were aroused, and he pursued it —

"Will they scourge thee? No, no, they would not be so cruel! Say they would not. Come, they *will* not, will they?"

The women betrayed confusion and distress, but there was no

avoiding an answer, so one of them said, in a voice choked with emotion—

"O, thou'lt break our hearts, thou gentle spirit!—God will help us to bear our"—

"It is a confession!" the king broke in. "Then they *will* scourge thee, the stonyhearted wretches! But O, thou must not weep, I cannot bear it. Keep up thy courage—I shall come to my own in time to save thee from this bitter thing, and I will do it!"

When the king awoke in the morning, the women were gone.

"They are saved!" he said, joyfully; then added, despondently, "but woe is me!—for they were my comforters."

Each of them had left a shred of ribbon pinned to his clothing, in token of remembrance. He said he would keep these things always; and that soon he would seek out these dear good friends of his and take them under his protection.

Just then the jailer came in with some subordinates and commanded that the prisoners be conducted to the jail-yard. The king was overjoyed—it would be a blessed thing to see the blue sky and breathe the fresh air once more. He fretted and chafed at the slowness of the officers, but his turn came at last and he was released from his staple and ordered to follow the other prisoners, with Hendon.

The court or quadrangle, was stone-paved, and open to the sky. The prisoners entered it through a massive archway of masonry, and were placed in file, standing, with their backs against the wall. A rope was stretched in front of them, and they were also guarded by their officers. It was a chill and lowering morning, and a light snow which had fallen during the night whitened the great empty space and added to the general dismalness of its aspect. Now and then a wintry wind shivered through the place and sent the snow eddying hither and thither.

In the centre of the court stood two women, chained to posts. A glance showed the king that these were his good friends. He shuddered, and said to himself, "Alack, they are not gone free, as I had thought. To think that such as these should know the lash!—in England! Ay there's the shame of it—not in Heathenesse, but Christian England! They will be scourged; and I, whom they have comforted and kindly entreated, must look on and see the great wrong done; it is strange, so strange! that I, the very source of power in this broad realm, am helpless to protect them. But let these miscreants look well to themselves, for there is a day coming when I will require of them a heavy reckoning for this work. For every blow they strike now, they shall feel a hundred, then."

"TWO WOMEN CHAINED TO POSTS."

A great gate swung open and a crowd of citizens poured in. They flocked around the two women, and hid them from the king's view. A clergyman entered and passed through the crowd, and he also was hidden. The king now heard talking, back and forth, as if questions were being asked and answered, but he could not make out what was said. Next there was a deal of bustle and preparation, and much passing and repassing of officials through that part of the crowd that stood on the further side of the women; and whilst this proceeded a deep hush gradually fell upon the people.

Now, by command, the masses parted and fell aside, and the king saw a spectacle that froze the marrow in his bones. Fagots had been piled about the two women, and a kneeling man was lighting them!

The women bowed their heads, and covered their faces with their hands; the yellow flames began to climb upward among the snapping and crackling fagots, and wreaths of blue smoke to stream away on the wind; the clergyman lifted his hands and began a prayer — just then two young girls came flying through the great gate, uttering piercing screams, and threw themselves upon the women at the stake. Instantly they were torn away by the officers, and one of them was kept in a tight grip, but the other broke loose, saying she would die with her mother; and before she could be stopped she had flung her arms about her mother's neck again. She was torn away once more, and with her gown on fire. Two or three men held her, and the burning portion of her gown was snatched off and thrown flaming aside, she struggling all the while to free herself, and saying she would be alone in the world, now, and begging to be allowed to die with her mother. Both the girls screamed continually, and fought for freedom; but suddenly this tumult was drowned under a volley of heart-piercing shrieks of mortal agony, — the king glanced from the frantic girls to the stake, then turned away and leaned his ashen face against the wall, and looked no more. He said, "That which I have seen, in that one little moment, will never go out from my memory, but will abide there; and I shall see it all the days, and dream of it all the nights, till I die. Would God I had been blind!"

Hendon was watching the king. He said to himself, with satisfaction, "His disorder mendeth; he hath changed, and groweth gentler. If he had followed his wont, he would have stormed at these varlets, and said he was king, and commanded that the women be turned loose unscathed. Soon his delusion will pass away and be forgotten, and his poor mind will be whole again. God speed the day!"

That same day several prisoners were brought in to remain over night, who were being conveyed, under guard, to various places in the kingdom, to undergo punishment for crimes committed. The king conversed with these,—he had made it a point, from the beginning,

"TORN AWAY BY THE OFFICERS."

to instruct himself for the kingly office by questioning prisoners whenever the opportunity offered—and the tale of their woes wrung his heart. One of them was a poor half-witted woman who had stolen a yard or two of cloth from a weaver—she was to be hanged for it.

Another was a man who had been accused of stealing a horse; he said the proof had failed, and he had imagined that he was safe from the halter; but no — he was hardly free before he was arraigned for killing a deer in the king's park; this was proved against him, and now he was on his way to the gallows. There was a tradesman's apprentice whose case particularly distressed the king; this youth said he found a hawk, one evening, that had escaped from its owner, and he took it home with him, imagining himself entitled to it; but the court convicted him of stealing it, and sentenced him to death.

"THE KING WAS FURIOUS."

The king was furious over these inhumanities, and wanted Hendon to break jail and fly with him to Westminster, so that he could mount his throne and hold out his sceptre in mercy over these unfortunate people and save their lives. "Poor child," sighed Hendon, "these woful tales have brought his malady upon him again — alack, but for this evil hap, he would have been well in a little time."

Among these prisoners was an old lawyer — a man with a strong face and a dauntless mien. Three years past, he had written a pamphlet against the Lord Chancellor, accusing him of injustice, and had been punished for it by the loss of his ears in the pillory, and degradation from the bar, and in addition had been fined £3000 and sentenced to imprisonment for life. Lately he had repeated his offence; and in consequence was now under sentence to lose *what remained of his ears*, pay a fine of £5000, be branded on both cheeks, and remain in prison for life.

"These be honorable scars," he said, and turned back his gray hair and showed the mutilated stubs of what had once been his ears.

The king's eye burned with passion. He said—

"None believe in me—neither wilt thou. But no matter—within the compass of a month thou shalt be free; and more, the laws that have dishonored thee, and shamed the English name, shall be swept from the statute books. The world is made wrong, kings should go to school to their own laws, at times, and so learn mercy." [1]

[1] See Notes to Chapter 27, at end of volume.

THE SACRIFICE

CHAPTER XXVIII.

THE SACRIFICE.

MEANTIME Miles was growing sufficiently tired of confinement and inaction. But now his trial came on, to his great gratification, and he thought he could welcome any sentence provided a further imprisonment should not be a part of it. But he was mistaken about that. He was in a fine fury when he found himself described as a "sturdy vagabond" and sentenced to sit two hours in the pillory for bearing that character and for assaulting the master of Hendon Hall. His pretensions as to brothership with his prosecutor, and rightful heirship to the Hendon honors and estates, were left contemptuously unnoticed, as being not even worth examination.

He raged and threatened, on his way to punishment, but it did no good; he was snatched roughly along, by the officers, and got an occasional cuff, besides, for his unreverent conduct.

The king could not pierce through the rabble that swarmed behind; so he was obliged to follow in the rear, remote from his good friend and servant. The king had been nearly condemned to the stocks, himself, for being in such bad company, but had been let off with a lecture and a warning, in consideration of his youth. When the crowd at last halted, he flitted feverishly from point to point around its outer rim, hunting a place to get through; and at last, after a deal of difficulty and delay, succeeded. There sat his poor henchman in the degrading stocks, the sport and butt of a dirty mob

—he, the body servant of the king of England! Edward had heard the sentence pronounced, but he had not realized the half that it meant. His anger began to rise as the sense of this new indignity which had been put upon him sank home; it jumped to summer heat, the next moment, when he saw an egg sail through the air and crush itself against Hendon's cheek, and heard the

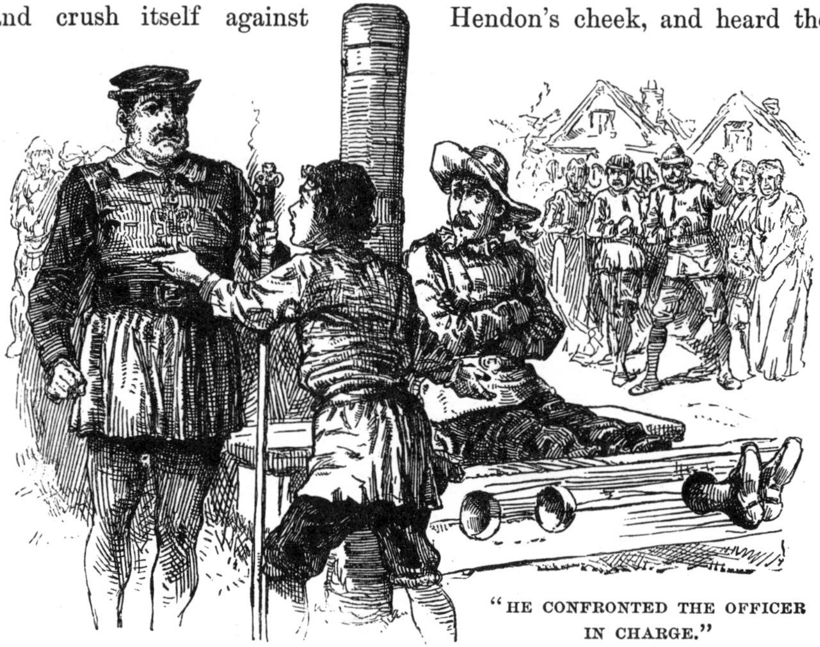

"HE CONFRONTED THE OFFICER IN CHARGE."

crowd roar its enjoyment of the episode. He sprang across the open circle and confronted the officer in charge, crying"—

"For shame! This is my servant—set him free! I am the"—

"O, peace!" exclaimed Hendon, in a panic, "thou'lt destroy thyself. Mind him not, officer, he is mad."

"Give thyself no trouble as to the matter of minding him, good man, I have small mind to mind him; but as to teaching him somewhat, to that I am well inclined." He turned to a subordinate and said, "Give the little fool a taste or two of the lash, to mend his manners."

"Half a dozen will better serve his turn," suggested Sir Hugh, who had ridden up, a moment before, to take a passing glance at the proceedings.

The king was seized. He did not even struggle, so paralyzed was he with the mere thought of the monstrous outrage that was proposed to be inflicted upon his sacred person. History was already defiled with the record of the scourging of an English king with whips — it was an intolerable reflection that he must furnish a duplicate of that shameful page. He was in the toils, there was no help for him: he must either take this punishment or beg for its remission. Hard conditions; he would take the stripes — a king might do that, but a king could not beg.

But meantime, Miles Hendon was resolving the difficulty. "Let the child go," said he; "ye heartless dogs, do ye not see how young and frail he is? Let him go — I will take his lashes."

"Marry, a good thought, — and thanks for it," said Sir Hugh, his face lighting with a sardonic satisfaction. "Let the little beggar go, and give this fellow a dozen in his place — an honest dozen, well laid on." The king was in the act of entering a fierce protest, but Sir Hugh silenced him with the potent remark, "Yes, speak up, do, and free thy mind — only, mark ye, that for each word you utter he shall get six strokes the more."

Hendon was removed from the stocks, and his back laid bare; and whilst the lash was applied the poor little king turned away his face and allowed unroyal tears to channel his cheeks unchecked. "Ah, brave good heart," he said to himself, "this loyal deed shall never perish out of my memory. I will not forget it — and neither shall *they!*" he added, with passion. Whilst he mused, his appreciation of Hendon's magnanimous conduct grew to greater and still greater dimensions in his mind, and so also did his gratefulness for it. Presently he said to himself, "Who saves his prince from wounds and possible death — and this he did for me — performs high service; but it is little — it is nothing! — O, less than nothing! — when 'tis weighed against the act of him who saves his prince from SHAME!"

"WHILE THE LASH WAS APPLIED, THE POOR KING TURNED AWAY HIS FACE."

Hendon made no outcry, under the scourge, but bore the heavy blows with soldierly fortitude. This, together with his redeeming the boy by taking his stripes for him, compelled the respect of even that forlorn and degraded mob that was gathered there; and its gibes and hootings died away, and no sound remained but the sound of the falling blows. The stillness that pervaded the place, when Hendon found himself once more in the stocks, was in strong contrast with the insulting clamor which had prevailed there so little a while before. The king came softly to Hendon's side, and whispered in his ear —

"Kings cannot ennoble thee, thou good, great soul, for One who is higher than kings hath done that for thee; but a king can confirm thy nobility to men." He picked up the scourge from the ground, touched Hendon's bleeding shoulders lightly with it, and whispered, "Edward of England dubs thee earl!"

"SIR HUGH SPURRED AWAY."

Hendon was touched. The water welled to his eyes, yet at the same time the grisly humor of the situation and circumstances so undermined his gravity that it was all he could do to keep some sign of his inward mirth from showing outside. To be suddenly hoisted, naked and gory, from the common stocks to the Alpine altitude and splendor of an Earldom, seemed to him the last possibility in the line of the grotesque. He said to himself, "Now am I finely tinselled, indeed! The spectre-knight of the Kingdom of Dreams and Shadows is become a spectre-earl! — a dizzy flight for a callow wing! An' this go on, I shall presently be hung like a very may-pole with fantastic gauds and make-believe honors. But I shall value them, all valueless as they are, for the love that doth bestow them. Better these poor mock dignities of mine, that come unasked, from a clean hand and a right spirit, than real ones bought by servility from grudging and interested power."

The dreaded Sir Hugh wheeled his horse about, and as he spurred away, the living wall divided silently to let him pass, and as silently closed together again. And so remained; nobody went so far as to venture a remark in favor of the prisoner, or in compliment to him; but no matter, the absence of abuse was a sufficient homage in itself. A late comer who was not posted as to the present circumstances, and who delivered a sneer at the "impostor" and was in the act of following it with a dead cat, was promptly knocked down and kicked out, without any words, and then the deep quiet resumed sway once more.

To London

CHAPTER XXIX.

TO LONDON.

WHEN Hendon's term of service in the stocks was finished, he was released and ordered to quit the region and come back no more. His sword was restored to him, and also his mule and his donkey. He mounted and rode off, followed by the king, the crowd opening with quiet respectfulness to let them pass, and then dispersing when they were gone.

Hendon was soon absorbed in thought. There were questions of high import to be answered. What should he do? Whither should he go? Powerful help must be found, somewhere, or he must relinquish his inheritance and remain under the imputation of being an impostor besides. Where could he hope to find this powerful help? Where, indeed! It was a knotty question. By and by a thought occurred to him which pointed to a possibility — the slenderest of slender possibilities, certainly, but still worth considering, for lack of any other that promised any thing at all. He remembered what old Andrews had said about the young king's goodness and his generous championship of the wronged and unfortunate. Why not go and try to get speech of him and beg for justice? Ah, yes, but could so fantastic a pauper get admission to the august presence of a monarch? Never mind — let that matter take care of itself; it was a bridge that would not need to be crossed till he should come to it. He was an old campaigner, and used to inventing shifts and expedients; no doubt he would be able to find a way. Yes, he would strike for the capital. Maybe his father's old friend Sir Humphrey Marlow would help him — "good old Sir Humphrey, Head Lieutenant of the late king's kitchen, or stables, or something" — Miles could not remember just what or which. Now that he had something to turn his energies to, a distinctly defined object

"HENDON MOUNTED AND RODE OFF WITH THE KING."

to accomplish, the fog of humiliation and depression which had settled down upon his spirits lifted and blew away, and he raised his head and looked about him. He was surprised to see how far he had come; the village was away behind him. The king was jogging along in his wake, with his head bowed; for he, too, was deep in plans and think-

"IN THE MIDST OF A JAM OF HOWLING PEOPLE."

ings. A sorrowful misgiving clouded Hendon's new-born cheerfulness: would the boy be willing to go again to a city where, during all his brief life, he had never known any thing but ill usage and pinching want? But the question must be asked; it could not be avoided; so Hendon reined up, and called out—

"I had forgotten to inquire whither we are bound. Thy commands, my liege!"

"To London!"

Hendon moved on again, mightily contented with the answer—but astounded at it, too.

The whole journey was made without an adventure of importance. But it ended with one. About ten o'clock on the night of the 19th of February, they stepped upon London Bridge, in the midst of a writhing, struggling jam of howling and hurrahing people, whose beer-jolly faces stood out strongly in the glare from manifold torches — and at that instant the decaying head of some former duke or other grandee tumbled down between them, striking Hendon on the elbow and then bounding off among the hurrying confusion of feet. So evanescent and unstable are men's works, in this world! — the late good king is but three weeks dead and three days in his grave, and already the adornments which he took such pains to select from prominent people for his noble bridge are falling. A citizen stumbled over that head, and drove his own head into the back of somebody in front of him, who turned and knocked down the first person that came handy, and was promptly laid out himself by that person's friend. It was the right ripe time for a free fight, for the festivities of the morrow — Coronation Day — were already beginning; everybody was full of strong drink and patriotism; within five minutes the free fight was occupying a good deal of ground; within ten or twelve it covered an acre or so, and was become a riot. By this time Hendon and the king were hopelessly separated from each other and lost in the rush and turmoil of the roaring masses of humanity. And so we leave them.

Tom's Progress

CHAPTER XXX.

TOM'S PROGRESS.

WHILST the true King wandered about the land poorly clad, poorly fed, cuffed and derided by tramps one while, herding with thieves and murderers in a jail another, and called idiot and impostor by all impartially, the mock King Tom Canty enjoyed a quite different experience.

When we saw him last, royalty was just beginning to have a bright side for him. This bright side went on brightening more and more every day: in a very little while it was become almost all sunshine and delightfulness. He lost his fears; his misgivings faded out and died; his embarrassments departed, and gave place to an easy and confident bearing. He worked the whipping-boy mine to ever-increasing profit.

He ordered my Lady Elizabeth and my Lady Jane Grey into his presence when he wanted to play or talk, and dismissed them when he was done with them, with the air of one familiarly accustomed to such performances. It no longer confused him to have these lofty personages kiss his hand at parting.

He came to enjoy being conducted to bed in state at night, and dressed with intricate and solemn ceremony in the morning. It came to be a proud pleasure to march to dinner attended by a glittering procession of officers of state and gentlemen-at-arms; insomuch, indeed, that he doubled his guard of gentlemen-at-arms, and made them a hundred. He liked to hear the bugles sounding down the long corridors, and the distant voices responding, "Way for the King!"

He even learned to enjoy sitting in throned state in council, and seeming to be something more than the Lord Protector's mouth-piece. He liked to receive great ambassadors and their gorgeous trains, and listen to the affectionate messages they brought from illustrious monarchs who called him "brother." O happy Tom Canty, late of Offal Court!

"TO KISS HIS HAND AT PARTING."

He enjoyed his splendid clothes, and ordered more: he found his four hundred servants too few for his proper grandeur, and trebled them. The adulation of salaaming courtiers came to be sweet music to his ears. He remained kind and gentle, and a sturdy and determined champion of all that were oppressed, and he made tireless war upon unjust laws: yet upon occasion, being offended, he could turn upon an earl, or even a duke, and give him a look that would make him tremble. Once, when his royal "sister," the grimly, holy Lady Mary, set herself to reason with him against the wisdom of his course in pardoning so many people who would otherwise be jailed, or hanged, or burned, and reminded him that their august late father's prisons had sometimes contained as high as sixty thousand convicts at one time,

"COMMANDED HER TO GO TO HER CLOSET."

and that during his admirable reign he had delivered seventy-two thousand thieves and robbers over to death by the executioner,[1] the boy was filled with generous indignation, and commanded her to go to her closet, and beseech God to take away the stone that was in her breast, and give her a human heart.

[1] Hume's England.

Did Tom Canty never feel troubled about the poor little rightful prince who had treated him so kindly, and flown out with such hot zeal to avenge him upon the insolent sentinel at the palace-gate? Yes; his first royal days and nights were pretty well sprinkled with painful thoughts about the lost prince, and with sincere longings for his return, and happy restoration to his native rights and splendors. But as time wore on, and the prince did not come, Tom's mind became more and more occupied with his new and enchanting experiences, and by little and little the vanished monarch faded almost out of his thoughts; and finally, when he did intrude upon them at intervals, he was become an unwelcome spectre, for he made Tom feel guilty and ashamed.

Tom's poor mother and sisters travelled the same road out of his mind. At first he pined for them, sorrowed for them, longed to see them, but later, the thought of their coming some day in their rags and dirt, and betraying him with their kisses, and pulling him down from his lofty place, and dragging him back to penury and degradation and the slums, made him shudder. At last they ceased to trouble his thoughts almost wholly. And he was content, even glad; for, whenever their mournful and accusing faces did rise before him now, they made him feel more despicable than the worms that crawl.

At midnight of the 19th of February, Tom Canty was sinking to sleep in his rich bed in the palace, guarded by his loyal vassals, and surrounded by the pomps of royalty, a happy boy; for to-morrow was the day appointed for his solemn crowning as King of England. At that same hour, Edward, the true king, hungry and thirsty, soiled and draggled, worn with travel, and clothed in rags and shreds, — his share of the results of the riot, — was wedged in among a crowd of people who were watching with deep interest certain hurrying gangs of workmen who streamed in and out of Westminster Abbey, busy as ants: they were making the last preparation for the royal coronation.

E VI
RECOGNITION
THE
PROCESSION

CHAPTER XXXI.

THE RECOGNITION PROCESSION.

WHEN Tom Canty awoke the next morning, the air was heavy with a thunderous murmur: all the distances were charged with it. It was music to him; for it meant that the English world was out in its strength to give loyal welcome to the great day.

Presently Tom found himself once more the chief figure in a wonderful floating pageant on the Thames; for by ancient custom the "recognition procession" through London must start from the Tower, and he was bound thither.

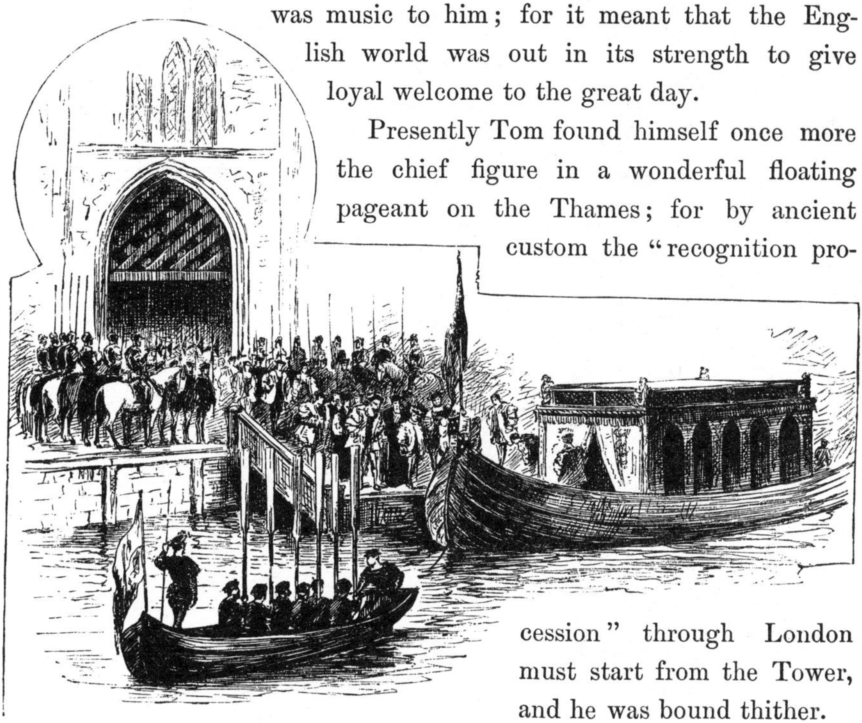

THE START FOR THE TOWER.

When he arrived there, the sides of the venerable fortress seemed suddenly rent in a thousand places, and from every rent leaped a red tongue of flame and a white gush of smoke; a deafening explosion followed, which drowned the shoutings of the multitude, and made the ground tremble; the flame-jets, the smoke, and the explosions, were repeated over and over again with marvellous celerity, so that in a few moments the old Tower disappeared in the vast fog of its own smoke, all but the very top of the tall pile called the White Tower: this, with its banners, stood out above the dense bank of vapor as a mountain-peak projects above a cloud-rack.

Tom Canty, splendidly arrayed, mounted a prancing war-steed, whose rich trappings almost reached to the ground; his "uncle," the Lord Protector Somerset, similarly mounted, took place in his rear; the King's Guard formed in single ranks on either side, clad in burnished armor; after the Protector followed a seemingly interminable procession of resplendent nobles attended by their vassals; after these came the lord mayor and the aldermanic body, in crimson velvet robes, and with their gold chains across their breasts; and after these the officers and members of all the guilds of London, in rich raiment, and bearing the showy banners of the several corporations. Also in the procession, as a special guard of honor through the city, was the Ancient and Honorable Artillery Company, — an organization already three hundred years old at that time, and the only military body in England possessing the privilege (which it still possesses in our day) of holding itself independent of the commands of Parliament. It was a brilliant spectacle, and was hailed with acclamations all along the line, as it took its stately way through the packed multitudes of citizens. The chronicler says, "The King, as he entered the city, was received by the people with prayers, welcomings, cries, and tender words, and all signs which argue an earnest love of subjects toward their sovereign; and the King, by holding up his glad countenance to such as stood afar off, and most tender language to those that stood nigh his Grace, showed himself no less thankful to receive the people's good will than they to offer it. To all that wished him well, he gave thanks. To such as bade 'God save his Grace,' he said in return, 'God save you all!' and added that 'he thanked them with all his heart.' Wonderfully transported were the people with the loving answers and gestures of their King."

In Fenchurch Street a "fair child, in costly apparel," stood on a stage to welcome his Majesty to the city. The last verse of his greeting was in these words:—

"Welcome, O King! as much as hearts can think;
Welcome again, as much as tongue can tell,—
Welcome to joyous tongues, and hearts that will not shrink:
God thee preserve, we pray, and wish thee ever well."

"WELCOME, O KING!"

The people burst forth in a glad shout, repeating with one voice what the child had said. Tom Canty gazed abroad over the surging sea of eager faces, and his heart swelled with exultation; and he felt that the one thing worth living for in this world was to be a king, and a nation's idol. Presently he caught sight, at a distance, of a couple of his ragged Offal Court comrades,—one of them the lord high admiral in his late mimic court, the other the first lord of the bedchamber in the same pretentious fiction; and his pride swelled higher than ever. Oh, if they could only recognize him now! What unspeakable glory it would be, if they could recognize him, and realize that the derided mock king of the slums and back alleys was become a real king, with illustrious dukes and princes for his humble menials, and the English world at his feet! But he had to deny himself, and

"A LARGESS! A LARGESS!"

choke down his desire, for such a recognition might cost more than it would come to: so he turned away his head, and left the two soiled lads to go on with their shoutings and glad adulations, unsuspicious of whom it was they were lavishing them upon.

Every now and then rose the cry, "A largess! a largess!" and Tom responded by scattering a handful of bright new coins abroad for the multitude to scramble for.

The chronicler says, "At the upper end of Gracechurch Street, before the sign of the Eagle, the city had erected a gorgeous arch, beneath which was a stage, which stretched from one side of the street to the other. This was a historical pageant, representing the King's immediate progenitors. There sat Elizabeth of York in the midst of an immense white rose, whose petals formed elaborate furbelows around her; by her side was Henry VII., issuing out of a vast red rose, disposed in the same manner: the hands of the royal pair were locked together, and the wedding-ring ostentatiously displayed. From the red and white roses proceeded a stem, which reached up to a second stage, occupied by Henry VIII., issuing from a red-and-white rose, with the effigy of the new king's mother, Jane Seymour, represented by his side. One branch sprang from this pair, which mounted to a third stage, where sat the effigy of Edward VI. himself, enthroned in royal majesty; and the whole pageant was framed with wreaths of roses, red and white."

This quaint and gaudy spectacle so wrought upon the rejoicing people, that their acclamations utterly smothered the small voice of the child whose business it was to explain the thing in eulogistic rhymes. But Tom Canty was not sorry; for this loyal uproar was sweeter music to him than any poetry, no matter what its quality might be. Whithersoever Tom turned his happy young face, the people recognized the exactness of his effigy's likeness to himself, the flesh and blood counterpart; and new whirlwinds of applause burst forth.

The great pageant moved on, and still on, under one triumphal arch after another, and past a bewildering succession of spectacular and symbolical tableaux, each of which typified and exalted some virtue, or talent, or merit, of the little king's. "Throughout the whole of Cheapside, from every penthouse and window, hung banners and streamers; and the richest carpets, stuffs, and cloth-of-gold tapestried the streets, — specimens of the great wealth of the stores within; and the splendor of this thoroughfare was equalled in the other streets, and in some even surpassed."

"And all these wonders and these marvels are to welcome me — me!" murmured Tom Canty.

The mock king's cheeks were flushed with excitement, his eyes were flashing, his senses swam in a delirium of pleasure. At this point, just as he was raising his hand to fling another rich largess, he caught sight of a pale, astounded face which was strained forward out of the second rank of the crowd, its intense eyes riveted upon him. A sickening consternation struck through him; he recognized his mother! and up flew his hand, palm outward, before his eyes, — that old involuntary gesture, born of a forgotten episode, and perpetuated by habit. In an instant more she had torn her way out of the press, and past the guards, and was at his side. She embraced his leg, she covered it with kisses, she cried, "O my child, my darling!" lifting toward him a face that was transfigured with joy and love. The same instant an officer of the King's Guard snatched her away with a curse, and sent her reeling back whence she came with a vigorous impulse from his strong arm. The words "I do not know you, woman!" were falling from Tom Canty's lips when this piteous thing occurred; but it smote him to the heart to see her treated so; and as she turned for a last glimpse of him, whilst the crowd was swallowing her from his sight, she seemed so wounded, so broken-hearted, that a shame fell upon him which consumed his pride to ashes, and withered his stolen royalty. His grandeurs were stricken valueless: they seemed to fall away from him like rotten rags.

The procession moved on, and still on, through ever augmenting splendors and ever augmenting tempests of welcome; but to Tom Canty they were as if they had not been. He neither saw nor heard.

"SHE WAS AT HIS SIDE."

Royalty had lost its grace and sweetness; its pomps were become a reproach. Remorse was eating his heart out. He said, "Would God I were free of my captivity!"

He had unconsciously dropped back into the phraseology of the first days of his compulsory greatness.

The shining pageant still went winding like a radiant and interminable serpent down the crooked lanes of the quaint old city, and through the huzzaing hosts; but still the King rode with bowed head and vacant eyes, seeing only his mother's face and that wounded look in it.

"Largess, largess!" The cry fell upon an unheeding ear.

"Long live Edward of England!" It seemed as if the earth shook with the explosion; but there was no response from the King. He heard it only as one hears the thunder of the surf when it is blown to the ear out of a great distance, for it was smothered under another sound which was still nearer, in his own breast, in his accusing conscience, — a voice which kept repeating those shameful words, "I do not know you, woman!"

The words smote upon the King's soul as the strokes of a funeral bell smite upon the soul of a surviving friend when they remind him of secret treacheries suffered at his hands by him that is gone.

New glories were unfolded at every turning; new wonders, new marvels, sprung into view; the pent clamors of waiting batteries were released; new raptures poured from the throats of the waiting multitudes: but the King gave no sign, and the accusing voice that went moaning through his comfortless breast was all the sound he heard.

By and by the gladness in the faces of the populace changed a little, and became touched with a something like solicitude or anxiety: an abatement in the volume of applause was observable too. The lord protector was quick to notice these things: he was as quick to detect the cause. He spurred to the King's side, bent low in his saddle, uncovered, and said, —

"My liege, it is an ill time for dreaming. The people observe thy downcast head, thy clouded mien, and they take it for an omen. Be advised: unveil the sun of royalty, and let it shine upon these boding vapors, and disperse them. Lift up thy face, and smile upon the people."

So saying, the duke scattered a handful of coins to right and left, then retired to his place. The mock king did mechanically as he had been bidden. His smile had no heart in it, but few eyes were near enough or sharp enough to detect that. The noddings of his plumed

"MY LIEGE, IT IS AN ILL TIME FOR DREAMING."

head as he saluted his subjects were full of grace and graciousness; the largess which he delivered from his hand was royally liberal: so the people's anxiety vanished, and the acclamations burst forth again in as mighty a volume as before.

Still once more, a little before the progress was ended, the duke was obliged to ride forward, and make remonstrance. He whispered,—

"O dread sovereign! shake off these fatal humors: the eyes of the world are upon thee." Then he added with sharp annoyance, "Perdition catch that crazy pauper! 'twas she that hath disturbed your Highness."

"SHE WAS MY MOTHER."

The gorgeous figure turned a lustreless eye upon the duke, and said in a dead voice,—

"She was my mother!"

"My God!" groaned the protector as he reined his horse backward to his post, "the omen was pregnant with prophecy. He is gone mad again!"

CORONATION-DAY

CHAPTER XXXII.

CORONATION DAY.

LET us go backward a few hours, and place ourselves in Westminster Abbey, at four o'clock in the morning of this memorable Coronation Day. We are not without company; for although it is still night, we find the torch-lighted galleries already filling up with people who are well content to sit still and wait seven or eight hours till the time shall come for them to see what they may not hope to see twice in their lives — the coronation of a king. Yes, London and Westminster have been astir ever since the warning guns boomed at three o'clock, and already crowds of untitled rich folk who have bought the privilege of trying to find sitting-room in the galleries are flocking in at the entrances reserved for their sort.

The hours drag along, tediously enough. All stir has ceased for some time, for every gallery has long ago been packed. We may sit, now, and look and think at our leisure. We have glimpses, here and there and yonder, through the dim cathedral twilight, of portions of many galleries and balconies, wedged full with people, the other portions of these galleries and balconies being cut off from sight by intervening pillars and architectural projections. We have in view the whole of the great north transept — empty, and waiting for England's privileged ones. We see also the ample area or platform, carpeted with rich stuffs, whereon the throne stands. The throne occupies the centre of the platform, and is raised above it upon an elevation of four steps. Within the seat of the throne is enclosed a rough flat rock — the stone of Scone — which many generations of Scottish kings sat on to be crowned, and so it in time became holy enough to answer a like purpose for English monarchs. Both the throne and its footstool are covered with cloth of gold.

Stillness reigns, the torches blink dully, the time drags heavily. But at last the lagging daylight asserts itself, the torches are extinguished, and a mellow radiance suffuses the great spaces. All features

"GATHERS UP THE LADY'S LONG TRAIN."

of the noble building are distinct, now, but soft and dreamy, for the sun is lightly veiled with clouds.

At seven o'clock the first break in the drowsy monotony occurs; for on the stroke of this hour the first peeress enters the transept, clothed like Solomon for splendor, and is conducted to her appointed place by an official clad in satins and velvets, whilst a duplicate of him gathers up the lady's long train, follows after, and, when the lady is seated, arranges the train across her lap for her. He then places her footstool according to her desire, after which he puts her coronet where it will be convenient to her hand when the time for the simultaneous coronetting of the nobles shall arrive.

By this time the peeresses are flowing in in a glittering stream, and the satin-clad officials are flitting and glinting everywhere, seating them and making them comfortable. The scene is animated enough, now. There is stir and life, and shifting color everywhere. After a time, quiet reigns again; for the peeresses are all come, and are all in their places — a solid acre, or such a matter, of human flowers, resplendent in variegated colors, and frosted like a Milky Way with diamonds. There are all ages, here: brown, wrinkled, whitehaired dowagers who are able to go back, and still back, down the stream of time, and recall the crowning of Richard III and the troublous days of that old forgotten age; and there are handsome middle-aged dames; and lovely and gracious young matrons; and gentle and beautiful young girls, with beaming eyes and fresh complexions, who may possibly put on their jewelled coronets awkwardly when the great time comes; for the matter will be new to them, and their excitement will be a sore hindrance. Still, this may not happen, for the hair of all these ladies has been arranged with a special view to the swift and successful lodging of the crown in its place when the signal comes.

We have seen that this massed array of peeresses is sown thick with diamonds, and we also see that it is a marvellous spectacle — but now we are about to be astonished in earnest. About nine, the clouds suddenly break away and a shaft of sunshine cleaves the mellow atmosphere, and drifts slowly along the ranks of ladies; and every rank it touches flames into a dazzling splendor of many-colored fires, and we tingle to our finger-tips with the electric thrill that is shot through us by the surprise and the beauty of the spectacle! Presently a special envoy from some distant corner of the Orient, marching with the general body of foreign ambassadors, crosses this bar of sunshine, and we catch our breath, the glory that streams and flashes and palpitates about him is so overpowering; for he is crusted from head to heel with gems, and his slightest movement showers a dancing radiance all around him.

Let us change the tense for convenience. The time drifted along, — one hour — two hours — two hours and a half; then the deep booming of artillery told that the king and his grand procession had arrived at last; so the waiting multitude rejoiced. All knew that a further delay must follow, for the king must be prepared and robed for the solemn ceremony; but this delay would be pleasantly occupied by the assembling of the peers of the realm in their stately robes. These were conducted ceremoniously to their seats, and their coronets placed conveniently at hand; and meanwhile the multitude in the galleries were alive with interest, for most of them were beholding for the first time, dukes, earls and barons, whose names had been historical for five hundred years. When all were finally seated, the spectacle from the galleries and all coigns of vantage was complete; a gorgeous one to look upon and to remember.

"TOM CANTY APPEARED."

Now the robed and mitred great heads of the church, and their attendants, filed in upon the platform and took their appointed places; these were followed by the Lord Protector and other great officials, and these again by a steel-clad detachment of the Guard.

There was a waiting pause; then, at a signal, a triumphant peal of music burst forth, and Tom Canty, clothed in a long robe of cloth of gold, appeared at a door, and stepped upon the platform. The entire multitude rose, and the ceremony of the Recognition ensued.

Then a noble anthem swept the Abbey with its rich waves of sound; and thus heralded and welcomed, Tom Canty was conducted to the throne. The ancient ceremonies went on, with impressive solemnity, whilst the audience gazed; and as they drew nearer and nearer to completion, Tom Canty grew pale, and still paler, and a deep and steadily deepening woe and despondency settled down upon his spirits and upon his remorseful heart.

At last the final act was at hand. The Archbishop of Canterbury lifted up the crown of England from its cushion and held it out over the trembling mock-king's head. In the same instant a rainbow-radiance flashed along the spacious transept; for with one impulse every individual in the great concourse of nobles lifted a coronet and poised it over his or her head, — and paused in that attitude.

A deep hush pervaded the Abbey. At this impressive moment, a startling apparition intruded upon the scene — an apparition observed by none in the absorbed multitude, until it suddenly appeared, moving up the great central aisle. It was a boy, bare-headed, ill shod, and clothed in coarse plebeian garments that were falling to rags. He raised his hand with a solemnity which ill comported with his soiled and sorry aspect, and delivered this note of warning —

"I forbid you to set the crown of England upon that forfeited head. *I* am the king!"

In an instant several indignant hands were laid upon the boy; but in the same instant Tom Canty, in his regal vestments, made a swift step forward and cried out in a ringing voice —

"Loose him and forbear! He *is* the king!"

A sort of panic of astonishment swept the assemblage, and they partly rose in their places and stared in a bewildered way at one

"AND FELL ON HIS KNEES BEFORE HIM."

another and at the chief figures in this scene, like persons who wondered whether they were awake and in their senses, or asleep and dreaming. The Lord Protector was as amazed as the rest, but quickly recovered himself and exclaimed in a voice of authority —

"Mind not his Majesty, his malady is upon him again — seize the vagabond!"

He would have been obeyed, but the mock-king stamped his foot and cried out —

"On your peril! Touch him not, he is the king!"

The hands were withheld; a paralysis fell upon the house; no one moved, no one spoke; indeed no one knew how to act or what to say, in so strange and surprising an emergency. While all minds were struggling to right themselves, the boy still moved steadily forward, with high port and confident mien; he had never halted from the beginning; and while the tangled minds still floundered helplessly, he stepped upon the platform, and the mock-king ran with a glad face to meet him; and fell on his knees before him and said —

"O, my lord the king, let poor Tom Canty be first to swear fealty to thee, and say 'Put on thy crown and enter into thine own again!'"

The Lord Protector's eye fell sternly upon the new-comer's face; but straightway the sternness vanished away, and gave place to an expression of wondering surprise. This thing happened also to the other great officers. They glanced at each other, and retreated a step by a common and unconscious impulse. The thought in each mind was the same: "What a strange resemblance!"

The Lord Protector reflected a moment or two, in perplexity, then he said, with grave respectfulness —

"By your favor, sir, I desire to ask certain questions which" —

"I will answer them, my lord."

The duke asked him many questions about the court, the late king, the prince, the princesses, — the boy answered them correctly and without hesitating. He described the rooms of state in the palace, the late king's apartments, and those of the Prince of Wales.

It was strange; it was wonderful; yes, it was unaccountable — so all said that heard it. The tide was beginning to turn, and Tom Canty's hopes to run high, when the Lord Protector shook his head and said —

"It is true it is most wonderful — but it is no more than our lord the king likewise can do." This remark, and this reference to himself as still the king, saddened Tom Canty, and he felt his hopes crumbling from under him. "These are not *proofs*," added the Protector.

The tide was turning very fast, now, very fast indeed — but in the wrong direction; it was leaving poor Tom Canty stranded on the throne, and sweeping the other out to sea. The Lord Protector communed with himself — shook his head — the thought forced itself upon him, "It is perilous to the State and to us all, to entertain so fateful a riddle as this; it could divide the nation and undermine the throne." He turned and said —

"Sir Thomas, arrest this — No, hold!" His face lighted, and he confronted the ragged candidate with this question —

"Where lieth the Great Seal? Answer me this truly, and the riddle is unriddled; for only he that was Prince of Wales *can* so answer! On so trivial a thing hang a throne and a dynasty!"

It was a lucky thought, a happy thought. That it was so considered by the great officials was manifested by the silent applause that shot from eye to eye around their circle in the form of bright approving glances. Yes, none but the true prince could dissolve the stubborn mystery of the vanished Great Seal — this forlorn little impostor had been taught his lesson well, but here his teachings must fail, for his teacher himself could not answer *that* question — ah, very good, very good indeed; now we shall be rid of this troublesome and perilous business in short order! And so they nodded invisibly and smiled inwardly with satisfaction, and looked to see this foolish lad stricken with a palsy of guilty confusion. How surprised they were, then, to see nothing of the sort happen — how they marvelled to hear him answer up promptly, in a confident and untroubled voice, and say —

"There is nought in this riddle that is difficult." Then, without so much as a by-your-leave to anybody, he turned and gave this command, with the easy manner of one accustomed to doing such things: "My lord St. John, go you to my private cabinet in the palace — for none knoweth the place better than you — and, close down to the floor, in the left corner remotest from the door that

"THE GREAT SEAL — FETCH IT HITHER."

opens from the ante-chamber, you shall find in the wall a brazen nail-head; press upon it and a little jewel-closet will fly open which not even you do know of — no, nor any soul else, in all the world but me and the trusty artisan that did contrive it for me. The first thing that falleth under your eye will be the Great Seal — fetch it hither."

All the company wondered at this speech, and wondered still more to see the little mendicant pick out this peer without hesitancy or apparent fear of mistake, and call him by name with such a placidly convincing air of having known him all his life. The peer was almost surprised into obeying. He even made a movement as if to go, but quickly recovered his tranquil attitude and confessed his blunder with a blush. Tom Canty turned upon him and said, sharply —

"Why dost thou hesitate? Hast not heard the king's command? Go!"

The lord St. John made a deep obeisance — and it was observed that it was a significantly cautious and non-committal one, it not being delivered at either of the kings, but at the neutral ground about half way between the two — and took his leave.

Now began a movement of the gorgeous particles of that official group which was slow, scarcely perceptible, and yet steady and persistent — a movement such as is observed in a kaleidoscope that is turned slowly, whereby the components of one splendid cluster fall away and join themselves to another — a movement which little by little, in the present case, dissolved the glittering crowd that stood about Tom Canty and clustered it together again in the neighborhood of the new-comer. Tom Canty stood almost alone. Now ensued a brief season of deep suspense and waiting — during which even the few faint-hearts still remaining near Tom Canty gradually scraped together courage enough to glide, one by one, over to the majority. So at last Tom Canty, in his royal robes and jewels, stood wholly alone and isolated from the world, a conspicuous figure, occupying an eloquent vacancy.

Now the lord St. John was seen returning. As he advanced up the mid-aisle the interest was so intense that the low murmur of conversation in the great assemblage died out and was succeeded by a profound hush, a breathless stillness, through which his footfalls pulsed with a dull and distant sound. Every eye was fastened upon him as he moved along. He reached the platform, paused a moment, then moved toward Tom Canty with a deep obeisance, and said —

"Sire, the Seal is not there!"

A mob does not melt away from the presence of a plague-patient

"SIRE, THE SEAL IS NOT THERE."

with more haste than the band of pallid and terrified courtiers melted away from the presence of the shabby little claimant of the Crown. In a moment he stood all alone, without friend or supporter, a target upon which was concentrated a bitter fire of scornful and angry looks. The Lord Protector called out fiercely —

"Cast the beggar into the street, and scourge him through the town — the paltry knave is worth no more consideration!"

Officers of the guard sprang forward to obey, but Tom Canty waved them off and said —

"Back! Whoso touches him perils his life!"

The Lord Protector was perplexed, in the last degree. He said to the lord St. John —

"Searched you well? — but it boots not to ask that. It doth seem passing strange. Little things, trifles, slip out of one's ken, and one does not think it matter for surprise; but how a so bulky thing as the Seal of England can vanish away and no man be able to get track of it again — a massy golden disk" —

Tom Canty, with beaming eyes, sprang forward and shouted —

"Hold, that is enough! Was it round? — and thick? — and had it letters and devices graved upon it? — Yes? O, *now* I know what this Great Seal is that there's been such worry and pother about! An' ye had described it to me, ye could have had it three weeks ago. Right well I know where it lies; but it was not I that put it there — first."

"Who, then, my liege?" asked the Lord Protector.

"He that stands there — the rightful King of England. And he shall tell you himself where it lies — then you will believe he knew it of his own knowledge. Bethink thee, my king — spur thy memory — it was the last, the very *last* thing thou didst that day before thou didst rush forth from the palace, clothed in my rags, to punish the soldier that insulted me."

A silence ensued, undisturbed by a movement or a whisper, and all eyes were fixed upon the new-comer, who stood, with bent head and corrugated brow, groping in his memory among a thronging multitude of valueless recollections for one single little elusive fact, which, found, would seat him upon a throne — unfound, would leave him as he was, for good and all — a pauper and an outcast. Moment after moment passed — the moments built themselves into minutes — still the boy struggled silently on, and gave no sign. But at last he heaved a sigh, shook his head slowly, and said, with a trembling lip and in a despondent voice —

"I call the scene back — all of it — but the Seal hath no place in it." He paused, then looked up, and said with gentle dignity, "My lords and gentlemen, if ye will rob your rightful sovereign of his own for lack of this evidence which he is not able to furnish, I may not stay ye, being powerless. But"—

"O, folly, O, madness, my king!" cried Tom Canty, in a panic, "wait!—think! Do not give up!—the cause is not lost! Nor *shall*

"BETHINK THEE, MY KING."

be, neither! List to what I say—follow every word—I am going to bring that morning back again, every hap just as it happened. We talked—I told you of my sisters, Nan and Bet—ah, yes, you remember that; and about mine old grandam—and the rough games of the lads of Offal Court—yes, you remember these things also; very well, follow me still, you shall recall every thing. You gave me food and drink, and did with princely courtesy send away the servants, so that my low breeding might not shame me before them—ah, yes, this also you remember."

As Tom checked off his details, and the other boy nodded his head in recognition of them, the great audience and the officials stared in puzzled wonderment; the tale sounded like true history, yet how could this impossible conjunction between a prince and a beggar boy have come about? Never was a company of people so perplexed, so interested, and so stupefied, before.

"For a jest, my prince, we did exchange garments. Then we stood before a mirror; and so alike were we that both said it seemed as if there had been no change made — yes, you remember that. Then you noticed that the soldier had hurt my hand — look! here it is, I cannot yet even write with it, the fingers are so stiff. At this your Highness sprang up, vowing vengeance upon that soldier, and ran toward the door — you passed a table — that thing you call the Seal lay on that table — you snatched it up and looked eagerly about, as if for a place to hide it — your eye caught sight of" —

"There, 'tis sufficient! — and the dear God be thanked!" exclaimed the ragged claimant, in a mighty excitement. "Go, my good St. John, — in an arm-piece of the Milanese armor that hangs on the wall, thou'lt find the Seal!"

"Right, my king! right!" cried Tom Canty; "*now* the sceptre of England is thine own; and it were better for him that would dispute it that he had been born dumb! Go, my lord St. John, give thy feet wings!"

The whole assemblage was on its feet, now, and well nigh out of its mind with uneasiness, apprehension, and consuming excitement. On the floor and on the platform a deafening buzz of frantic conversation burst forth, and for some time nobody knew any thing or heard any thing or was interested in any thing but what his neighbor was shouting into his ear, or he was shouting into his neighbor's ear. Time — nobody knew how much of it — swept by unheeded and unnoted. — At last a sudden hush fell upon the house, and in the same moment

St. John appeared upon the platform and held the Great Seal aloft in his hand. Then such a shout went up!

"Long live the true King!"

For five minutes the air quaked with shouts and the crash of musical instruments, and was white with a storm of waving handkerchiefs; and through it all a ragged lad, the most conspicuous figure in England, stood, flushed and happy and proud, in the centre of the spacious platform, with the great vassals of the kingdom kneeling around him.

"LONG LIVE THE TRUE KING!"

Then all rose, and Tom Canty cried out—

"Now, O, my king, take these regal garments back, and give poor Tom, thy servant, his shreds and remnants again."

The Lord Protector spoke up—

"Let the small varlet be stripped and flung into the Tower."

But the new king, the true king, said—

"I will not have it so. But for him I had not got my crown again—none shall lay a hand upon him to harm him. And as for thee, my good uncle, my Lord Protector, this conduct of thine is not grateful toward this poor lad, for I hear he hath made thee a duke"—the Protector blushed—"yet he was not a king; wherefore, what is thy fine title worth, now? To-morrow you shall sue to me, *through him*, for its confirmation, else no duke, but a simple earl, shalt thou remain."

Under this rebuke, his grace the duke of Somerset retired a little from the front for the moment. The king turned to Tom, and said, kindly—

"My poor boy, how was it that you could remember where I hid the Seal when I could not remember it myself?"

"Ah, my king, that was easy, since I used it divers days."

"Used it,—yet could not explain where it was?"

"I did not know it was *that* they wanted. They did not describe it, your majesty."

"Then how used you it?"

The red blood began to steal up into Tom's cheeks, and he dropped his eyes and was silent.

"Speak up, good lad, and fear nothing," said the king. "How used you the Great Seal of England?"

Tom stammered a moment, in a pathetic confusion, then got it out—

"To crack nuts with!"

Poor child, the avalanche of laughter that greeted this, nearly swept him off his feet. But if a doubt remained in any mind that Tom Canty was not the king of England and familiar with the august appurtenances of royalty, this reply disposed of it utterly.

Meantime the sumptuous robe of state had been removed from Tom's shoulders to the king's, whose rags were effectually hidden from sight under it. Then the coronation ceremonies were resumed;

"TO CRACK NUTS WITH."

the true king was anointed and the crown set upon his head, whilst cannon thundered the news to the city, and all London seemed to rock with applause.

EDUARD
AS
KING

CHAPTER XXXIII.

EDWARD AS KING.

MILES HENDON was picturesque enough before he got into the riot on London Bridge — he was more so when he got out of it. He had but little money when he got in, none at all when he got out. The pickpockets had stripped him of his last farthing.

But no matter, so he found his boy. Being a soldier, he did not go at his task in a random way, but set to work, first of all, to arrange his campaign.

What would the boy naturally do? Where would he naturally go? Well — argued Miles — he would naturally go to his former haunts, for that is the instinct of unsound minds, when homeless and forsaken, as well as of sound ones. Whereabouts were his former haunts? His rags, taken together with the low villain who seemed to know him and who even claimed to be his father, indicated that his home was in one or another of the poorest and meanest districts of London. Would the search for him be difficult, or long? No, it was likely to be easy and brief. He would not hunt for the boy, he would hunt for a crowd; in the centre of a big crowd or a little one, sooner or later, he should find his poor little friend, sure; and the mangy mob would be entertaining itself with pestering and aggravating the boy, who would be proclaiming himself king, as usual. Then Miles Hendon would cripple some of those people, and carry off his little ward, and comfort and cheer him with loving words, and the two would never be separated any more.

So Miles started on his quest. Hour after hour he tramped through back alleys and squalid streets, seeking groups and crowds, and finding no end of them, but never any sign of the boy. This

"HE STRETCHED HIMSELF ON THE GROUND."

greatly surprised him, but did not discourage him. To his notion, there was nothing the matter with his plan of campaign; the only miscalculation about it was that the campaign was becoming a lengthy one, whereas he had expected it to be short.

When daylight arrived, at last, he had made many a mile, and canvassed many a crowd, but the only result was that he was tolerably tired, rather hungry, and very sleepy. He wanted some breakfast, but there was no way to get it. To beg for it did not occur to him; as to pawning his sword, he would as soon have thought of parting with his honor; he could spare some of his clothes — yes, but one could as easily find a customer for a disease as for such clothes.

At noon he was still tramping — among the rabble which followed after the royal procession, now; for he argued that this regal display would attract his little lunatic powerfully. He followed the pageant through all its devious windings about London, and all the way to Westminster and the Abbey. He drifted here and there amongst the multitudes that were massed in the vicinity for a weary long time, baffled and perplexed, and finally wandered off, thinking, and trying to contrive some way to better his plan of campaign. By and by, when he came to himself out of his musings, he discovered that the town was far behind him and that the day was growing old. He was near the river, and in the country; it was a region of fine rural seats — not the sort of district to welcome clothes like his.

It was not at all cold; so he stretched himself on the ground in the lee of a hedge to rest and think. Drowsiness presently began to settle upon his senses; the faint and far-off boom of cannon was wafted to his ear, and he said to himself "The new king is crowned," and straightway fell asleep. He had not slept or rested, before, for more than thirty hours. He did not wake again until near the middle of the next morning.

He got up, lame, stiff, and half famished, washed himself in the river, stayed his stomach with a pint or two of water, and trudged off toward Westminster grumbling at himself for having wasted so much time. Hunger helped him to a new plan, now; he would try to get speech with old Sir Humphrey Marlow and borrow a few marks, and — but that was enough of a plan for the present; it would be time enough to enlarge it when this first stage should be accomplished.

Toward eleven o'clock he approached the palace; and although a host of showy people were about him, moving in the same direction, he was not inconspicuous — his costume took care of that. He watched these people's faces narrowly, hoping to find a charitable one whose possessor might be willing to carry his name to the old lieutenant — as to trying to get into the palace himself, that was simply out of the question.

Presently our whipping-boy passed him, then wheeled about and scanned his figure well, saying to himself, "An' that is not the very vagabond his majesty is in such a worry about, then am I an ass — though belike I was that before. He answereth the description to a rag — that God should make two such, would be to cheapen miracles, by wasteful repetition. I would I could contrive an excuse to speak with him."

Miles Hendon saved him the trouble; for he turned about, then, as a man generally will when somebody mesmerizes him by gazing hard at him from behind; and observing a strong interest in the boy's eyes, he stepped toward him and said —

"You have just come out from the palace; do you belong there?"

"Yes, your worship."

"Know you Sir Humphrey Marlow?"

The boy started, and said to himself, "Lord! mine old departed father!" Then he answered, aloud, "Right well, your worship."

"Good — is he within?"

"Yes," said the boy; and added, to himself, "within his grave."

"Might I crave your favor to carry my name to him, and say I beg to say a word in his ear?"

"I will despatch the business right willingly, fair sir."

"Then say Miles Hendon, son of Sir Richard, is here without — I shall be greatly bounden to you, my good lad."

The boy looked disappointed — "the king did not name him so," he said to himself — "but it mattereth not, this is his twin brother, and can give his majesty news of 'tother Sir-Odds-and-Ends, I warrant." So he said to Miles, "Step in there a moment, good sir, and wait till I bring you word."

Hendon retired to the place indicated — it was a recess sunk in the palace wall, with a stone bench in it — a shelter for sentinels in bad weather. He had hardly seated himself when some halberdiers, in charge of an officer, passed by. The officer saw him, halted his

"ARRESTED AS A SUSPICIOUS CHARACTER."

men, and commanded Hendon to come forth. He obeyed, and was promptly arrested as a suspicious character prowling within the precincts of the palace. Things began to look ugly. Poor Miles was going to explain, but the officer roughly silenced him, and ordered his men to disarm him and search him.

"God of his mercy grant that they find somewhat," said poor Miles; "I have searched enow, and failed, yet is my need greater than theirs."

Nothing was found but a document. The officer tore it open, and Hendon smiled when he recognized the "pot-hooks" made by his lost little friend that black day at Hendon Hall. The officer's face grew dark as he read the English paragraph, and Miles blenched to the opposite color as he listened.

"Another new claimant of the crown!" cried the officer. "Verily they breed like rabbits, to-day. Seize the rascal, men, and see ye keep him fast whilst I convey this precious paper within and send it to the king."

He hurried away, leaving the prisoner in the grip of the halberdiers.

"Now is my evil luck ended at last," muttered Hendon, "for I shall dangle at a rope's end for a certainty, by reason of that bit of writing. And what will become of my poor lad!—ah, only the good God knoweth."

By and by he saw the officer coming again, in a great hurry; so he plucked his courage together, purposing to meet his trouble as became a man. The officer ordered the men to loose the prisoner and return his sword to him; then bowed respectfully, and said—

"Please you sir, to follow me."

Hendon followed, saying to himself, "An' I were not travelling to death and judgment, and so must needs economize in sin, I would throttle this knave for his mock courtesy."

The two traversed a populous court, and arrived at the grand entrance of the palace, where the officer, with another bow, delivered Hendon into the hands of a gorgeous official, who received him with profound respect and led him forward through a great hall, lined on both sides with rows of splendid flunkies (who made reverential obeisance as the two passed along, but fell into death-throes of silent laughter at our stately scare-crow the moment his back was turned,) and up a broad staircase, among flocks of fine folk, and finally conducted him into a vast room, clove a passage for him through the assembled nobility of England, then made a bow, reminded him to take his hat off, and left him standing in the middle of the room, a mark for all eyes, for plenty of indignant frowns, and for a sufficiency of amused and derisive smiles.

Miles Hendon was entirely bewildered. There sat the young king, under a canopy of state, five steps away, with his head bent down and aside, speaking with a sort of human bird of paradise — a duke, maybe; Hendon observed to himself that it was hard enough to be sentenced to death in the full vigor of life, without having this peculiarly public humiliation added. He wished the king would hurry about it — some of the gaudy people near by were becoming pretty offensive. At this moment the king raised his head slightly and Hendon caught a good view of his face. The sight nearly took his breath away! — He stood gazing at the fair young face like one transfixed; then presently ejaculated —

"Lo, the lord of the Kingdom of Dreams and Shadows on his throne!"

He muttered some broken sentences, still gazing and marvelling; then turned his eyes around and about, scanning the gorgeous throng and the splendid saloon, murmuring "But these are *real* — verily these are *real* — surely it is not a dream."

He stared at the king again — and thought, "*Is* it a dream? . . . or *is* he the veritable sovereign of England, and not the friendless poor Tom o' Bedlam I took him for — who shall solve me this riddle?"

A sudden idea flashed in his eye, and he strode to the wall, gathered up a chair, brought it back, planted it on the floor, and sat down in it!

A buzz of indignation broke out, a rough hand was laid upon him, and a voice exclaimed, —

"Up, thou mannerless clown! — wouldst sit in the presence of the king?"

The disturbance attracted his majesty's attention, who stretched forth his hand and cried out —

"Touch him not, it is his right!"

"IT IS HIS RIGHT."

The throng fell back, stupefied. The king went on —

"Learn ye all, ladies, lords and gentlemen, that this is my trusty and well beloved servant, Miles Hendon, who interposed his good sword and saved his prince from bodily harm and possible death — and for this he is a knight, by the king's voice. Also learn, that for a higher service, in that he saved his sovereign stripes and shame, taking these upon himself, he is a peer of England,

Earl of Kent, and shall have gold and lands meet for the dignity. More — the privilege which he hath just exercised is his by royal grant; for we have ordained that the chiefs of his line shall have and hold the right to sit in the presence of the majesty of England henceforth, age after age, so long as the crown shall endure. Molest him not."

Two persons, who, through delay, had only arrived from the country during this morning, and had now been in this room only five minutes, stood listening to these words and looking at the king, then at the scare-crow, then at the king again, in a sort of torpid bewilderment. These were Sir Hugh and the Lady Edith. But the new Earl did not see them. He was still staring at the monarch, in a dazed way, and muttering —

"O, body o' me! *This* my pauper! This my lunatic! This is he whom *I* would show what grandeur was, in my house of seventy rooms and seven and twenty servants! This is he who had never known aught but rags for raiment, kicks for comfort, and offal for diet! This is he whom *I* adopted and would make respectable! Would God I had a bag to hide my head in!"

Then his manners suddenly came back to him, and he dropped upon his knees, with his hands between the king's, and swore allegiance and did homage for his lands and titles. Then he rose and stood respectfully aside, a mark still for all eyes — and much envy, too.

Now the king discovered Sir Hugh, and spoke out, with wrathful voice and kindling eye —

"Strip this robber of his false show and stolen estates, and put him under lock and key till I have need of him."

The late Sir Hugh was led away.

There was a stir at the other end of the room, now; the assemblage fell apart, and Tom Canty, quaintly but richly clothed, marched down, between these living walls, preceded by an usher. He knelt before the king, who said—

"I have learned the story of these past few weeks, and am well pleased with thee. Thou hast governed the realm with right royal gentleness and mercy. Thou hast found thy mother and thy sisters again? Good; they shall be cared for—and thy father shall

"STRIP THIS ROBBER."

hang, if thou desire it and the law consent. Know, all ye that hear my voice, that from this day, they that abide in the shelter of Christ's Hospital and share the king's bounty, shall have their minds and hearts fed, as well as their baser parts; and this boy shall dwell there, and hold the chief place in its honorable body of governors, during life. And for that he hath been a king, it is meet that other than common observance shall be his due; wherefore, note this his dress of state, for by it he shall be known, and none shall copy it; and wheresoever he shall come, it shall remind the people that he hath been royal, in his time, and none shall deny him his due of reverence or fail to give him salutation. He hath the

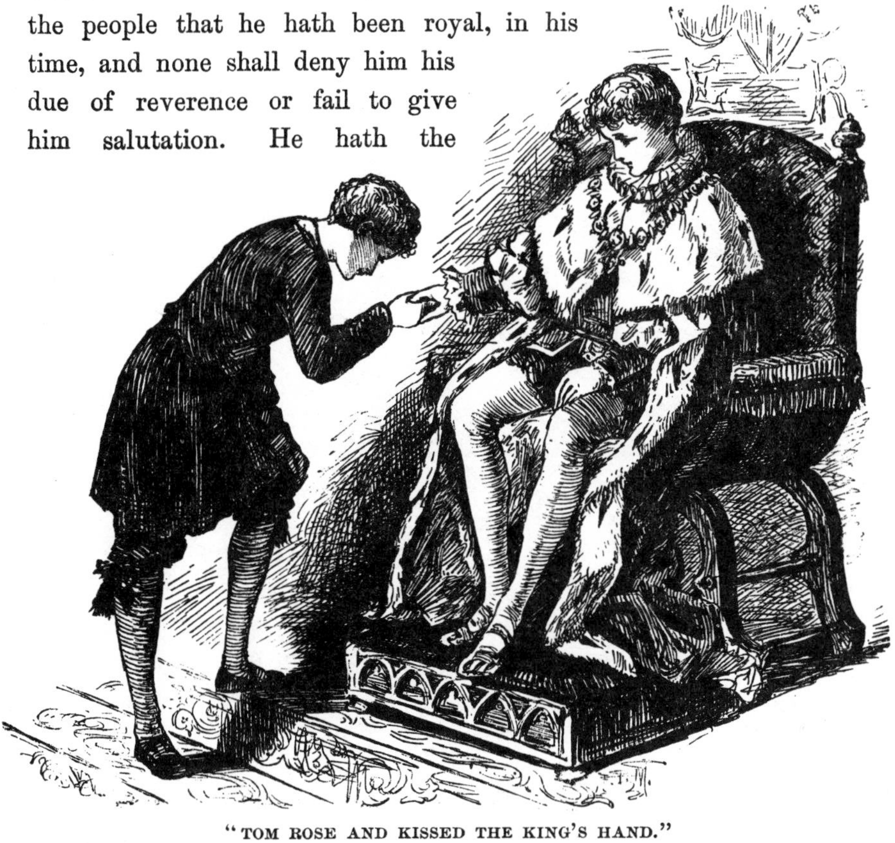

"TOM ROSE AND KISSED THE KING'S HAND."

throne's protection, he hath the crown's support, he shall be known and called by the honorable title of the King's Ward."

The proud and happy Tom Canty rose and kissed the king's hand, and was conducted from the presence. He did not waste any time, but flew to his mother, to tell her and Nan and Bet all about it and get them to help him enjoy the great news.[1]

[1] See Notes to Chapter 33 at end of the volume.

Justice and Retribution

CONCLUSION.

JUSTICE AND RETRIBUTION.

When the mysteries were all cleared up, it came out, by confession of Hugh Hendon, that his wife had repudiated Miles by his command, that day at Hendon Hall — a command assisted and supported by the perfectly trustworthy promise that if she did not deny that he was Miles Hendon, and stand firmly to it, he would have her life; whereupon she said take it, she did not value it — and she would not repudiate Miles; then the husband said he would spare her life but have Miles assassinated! This was a different matter; so she gave her word and kept it.

Hugh was not prosecuted for his threats or for stealing his brother's estates and title, because the wife and brother would not testify against him — and the former would not have been allowed to do it, even if she had wanted to. Hugh deserted his wife and went over to the continent, where he presently died; and by and by the earl of Kent married his relict. There were grand times and rejoicings at Hendon village when the couple paid their first visit to the Hall.

Tom Canty's father was never heard of again.

The king sought out the farmer who had been branded and sold as a slave, and reclaimed him from his evil life with the Ruffler's gang, and put him in the way of a comfortable livelihood.

He also took that old lawyer out of prison and remitted his fine. He provided good homes for the daughters of the two Baptist women whom he saw burned at the stake, and roundly punished the official who laid the undeserved stripes upon Miles Hendon's back.

He saved from the gallows the boy who had captured the stray falcon, and also the woman who had stolen a remnant of cloth from a weaver; but he was too late to save the man who had been convicted of killing a deer in the royal forest.

He showed favor to the justice who had pitied him when he was supposed to have stolen a pig, and he had the gratification of seeing him grow in the public esteem and become a great and honored man.

As long as the king lived he was fond of telling the story of his adventures, all through, from the hour that the sentinel cuffed him away from the palace gate till the final midnight when he deftly mixed himself into a gang of hurrying workmen and so slipped into the Abbey and climbed up and hid himself in the Confessor's tomb, and then slept so long, next day, that he came within one of missing the Coronation altogether. He said that the frequent rehearsing of the precious lesson kept him strong in his purpose to make its teachings yield benefits to his people; and so, whilst his life was spared he should continue to tell the story, and thus keep its sorrowful spectacles fresh in his memory and the springs of pity replenished in his heart.

Miles Hendon and Tom Canty were favorites of the king, all through his brief reign, and his sincere mourners when he died. The good earl of Kent had too much sense to abuse his peculiar privilege; but he exercised it twice after the instance we have seen of it before he was called from the world; once at the accession of Queen Mary, and once at the accession of Queen Elizabeth. A descendant of his exercised it at the accession of James I. Before this one's son chose to use the privilege, near a quarter of a century had elapsed, and the "privilege of the Kents" had faded out of most people's memories; so, when the Kent of that day appeared before Charles I and his court and sat down in the sovereign's presence to assert and perpetuate the right of his house, there was a fine stir, indeed! But the matter was soon explained, and the right confirmed. The last earl of the line fell in the wars of the Commonwealth fighting for the king, and the odd privilege ended with him.

Tom Canty lived to be a very old man, a handsome, white-haired old fellow, of grave and benignant aspect. As long as he lasted he was honored; and he was also reverenced, for his striking and peculiar costume kept the people reminded that "in his time he had been royal;" so, wherever he appeared the crowd fell apart, making way for him, and whispering, one to another, "Doff thy hat, it is the King's Ward!"—and so they saluted, and got his kindly smile in return—and they valued it, too, for his was an honorable history.

Yes, King Edward VI lived only a few years, poor boy, but he lived them worthily. More than once, when some great dignitary, some gilded vassal of the crown, made argument against his leniency, and urged that some law which he was bent upon amending was gentle enough for its purpose, and wrought no suffering or oppression which any one need mightily mind, the young king turned the mournful eloquence of his great compassionate eyes upon him and answered—

"What dost *thou* know of suffering and oppression? I and my people know, but not thou."

The reign of Edward VI was a singularly merciful one for those harsh times. Now that we are taking leave of him let us try to keep this in our minds, to his credit.

NOTES

NOTES.

Note 1.—Page 50.

Christ's Hospital Costume.

It is most reasonable to regard the dress as copied from the costume of the citizens of London of that period, when long blue coats were the common habit of apprentices and serving-men, and yellow stockings were generally worn; the coat fits closely to the body, but has loose sleeves, and beneath is worn a sleeveless yellow under-coat; around the waist is a red leathern girdle; a clerical band around the neck, and a small flat black cap, about the size of a saucer, completes the costume.—*Timbs' "Curiosities of London."*

Note 2.—Page 53.

It appears that Christ's Hospital was not originally founded as a *school;* its object was to rescue children from the streets, to shelter, feed, clothe them, etc.—*Timbs' "Curiosities of London."*

Note 3.—Page 67.

The Duke of Norfolk's Condemnation Commanded.

The King was now approaching fast towards his end; and fearing lest Norfolk should escape him, he sent a message to the Commons, by which he desired them to hasten the bill, on pretence that Norfolk enjoyed the dignity of earl marshal, and it was necessary to appoint another, who might officiate at the ensuing ceremony of installing his son Prince of Wales.—*Hume,* vol. iii. p. 307.

Note 4.—Page 91.

It was not till the end of this reign [Henry VIII] that any salads, carrots, turnips, or other edible roots were produced in England. The little of these vegetables that was used, was formerly imported from Holland and Flanders. Queen Catherine, when she wanted a salad, was obliged to despatch a messenger thither on purpose. — *Hume's History of England*, vol. iii. p. 314.

Note 5. — Page 100.

Attainder of Norfolk.

The house of peers, without examining the prisoner, without trial or evidence, passed a bill of attainder against him and sent it down to the commons. . . . The obsequious commons obeyed his [the King's] directions; and the King, having affixed the royal assent to the bill by commissioners, issued orders for the execution of Norfolk on the morning of the twenty-ninth of January, [the next day.] — *Hume's England*, vol. iii. p. 306.

Note 6. — Page 120.

The Loving-Cup.

The loving-cup, and the peculiar ceremonies observed in drinking from it, are older than English history. It is thought that both are Danish importations. As far back as knowledge goes, the loving-cup has always been drunk at English banquets. Tradition explains the ceremonies in this way: in the rude ancient times it was deemed a wise precaution to have both hands of both drinkers employed, lest while the pledger pledged his love and fidelity to the pledgee the pledgee take that opportunity to slip a dirk into him!

Note 7. — Page 129.

The Duke of Norfolk's Narrow Escape.

Had Henry VIII survived a few hours longer, his order for the duke's execution would have been carried into effect. "But news being carried to the Tower that the King himself had expired that night, the lieutenant deferred obeying the warrant; and it was not thought advisable by the Council to begin a new reign by the death of the greatest nobleman in the Kingdom, who had been condemned by a sentence so unjust and tyrannical." — *Hume's England*, vol. iii. p. 307.

Note 8. — Page 171.

The Whipping-Boy.

James I and Charles II had whipping-boys, when they were little fellows, to take their punishment for them when they fell short in their lessons; so I have ventured to furnish my small prince with one, for my own purposes.

Notes to Chapter XV. — Page 192.

Character of Hertford.

The young king discovered an extreme attachment to his uncle, who was, in the main, a man of moderation and probity. — *Hume's England*, vol. iii. p. 324.

But if he [the Protector] gave offence by assuming too much state, he deserves great praise on account of the laws passed this session, by which the rigor of former statutes was much mitigated, and some security given to the freedom of the constitution. All laws were repealed which extended the crime of treason beyond the statute of the twenty-fifth of Edward III; all laws enacted during the late reign extending the crime of felony; all the former laws against Lollardy or heresy, together with the statute of the Six Articles. None were to be accused for words, but within a month after they were spoken. By these repeals several of the most rigorous laws that ever had passed in England were annulled; and some dawn, both of civil and religious liberty, began to appear to the people. A repeal also passed of that law, the destruction of all laws, by which the king's proclamation was made of equal force with a statute. — *Ibid.*, vol. iii. p. 339.

Boiling to Death.

In the reign of Henry VIII, poisoners were, by act of parliament, condemned to be *boiled to death.* This act was repealed in the following reign.

In Germany, even in the 17th century, this horrible punishment was inflicted on coiners and counterfeiters. Taylor, the Water Poet, describes an execution he witnessed in Hamburg, in 1616. The judgment pronounced against a coiner of false money was that he should "be *boiled to death in oil;* not thrown into the vessel at once, but with a pulley or rope to be hanged under the armpits, and then let down into the oil *by degrees;* first the feet, and next the legs, and so to boil his flesh from his bones alive." — *Dr. J. Hammond Trumbull's "Blue Laws, True and False,"* p. 13.

The Famous Stocking Case.

A WOMAN and her daughter *nine years old*, were hanged in Huntingdon for selling their souls to the devil, and raising a storm by pulling off their stockings! — *Ibid.*, p. 20.

NOTE 10. — Page 214.

Enslaving.

So young a king, and so ignorant a peasant were likely to make mistakes — and this is an instance in point. This peasant was suffering from this law *by anticipation;* the king was venting his indignation against a law which was not yet in existence: for this hideous statute was to have birth in this little king's own reign However, we know, from the humanity of his character, that it could never have been suggested by him.

NOTES TO CHAPTER XXIII. — Page 285.

Death for Trifling Larcenies.

WHEN Connecticut and New Haven were framing their first codes, larceny above the value of twelve pence was a capital crime in England — as it had been since the time of Henry I. — *Dr. J. Hammond Trumbull's* "*Blue Laws, True and False,*" p. 17.

The curious old book called "The English Rogue" makes the limit thirteen pence ha'penny; death being the portion of any who steal a thing "above the value of thirteen pence ha'penny."

NOTES TO CHAPTER XXVII. — Page 317.

FROM many descriptions of larceny, the law expressly took away the benefit of clergy; to steal a horse, or a *hawk*, or woollen cloth from the weaver, was a hanging matter. So it was, to kill a deer from the king's forest, or to export sheep from the Kingdom. — *Dr. J. Hammond Trumbull's* "*Blue Laws, True and False,*" p. 13.

William Prynne, a learned barrister, was sentenced — [long after Edward the Sixth's time] — to lose both his ears in the pillory; to degradation from the bar; a fine of £3,000, and imprisonment for life. Three years afterwards, he gave new offence to Laud, by publishing a pamphlet against the hierarchy. He was again prosecuted, and was sentenced to lose *what remained of his ears;* to pay a fine of £5,000; to be *branded on both his cheeks* with the letters S. L. (for Seditious Libeller,) and to remain in prison for life. The severity of this sentence was equalled by the savage rigor of its execution. — *Ibid.*, p. 12.

Notes to Chapter XXXIII. —Page 395.

Christ's Hospital, or Blue Coat School, " the Noblest Institution in the World."

The ground on which the Priory of the Grey Friars stood was conferred by Henry the Eighth on the Corporation of London, [who caused the institution there of a home for poor boys and girls.] Subsequently, Edward the Sixth caused the old Priory to be properly repaired, and founded within it that noble establishment called the Blue Coat School, or Christ's Hospital, for the *education* and maintenance of orphans and the children of indigent persons. . . . Edward would not let him [Bishop Ridley] depart till the letter was written, [to the Lord Mayor,] and then charged him to deliver it himself, and signify his special request and commandment that no time might be lost in proposing what was convenient, and apprising him of the proceedings. The work was zealously undertaken, Ridley himself engaging in it; and the result was, the founding of Christ's Hospital for the Education of Poor Children. [The king endowed several other charities at the same time.] "Lord God," said he, "I yield thee most hearty thanks that thou hast given me life thus long, to finish this work to the glory of thy name!" That innocent and most exemplary life was drawing rapidly to its close, and in a few days he rendered up his spirit to his Creator, praying God to defend the realm from Papistry. — *J. Heneage Jesse's "London, its Celebrated Characters and Places."*

In the Great Hall hangs a large picture of King Edward VI seated on his throne, in a scarlet and ermined robe, holding the sceptre in his left hand, and presenting with the other the Charter to the kneeling Lord Mayor. By his side stands the Chancellor, holding the seals, and next to him are other officers of state. Bishop Ridley kneels before him with uplifted hands, as if supplicating a blessing on the event; whilst the Aldermen, etc., with the Lord Mayor, kneel on both sides, occupying the middle ground of the picture; and lastly, in front, are a double row of boys on one side, and girls on the other, from the master and matron down to the boy and girl who have stepped forward from their respective rows, and kneel with raised hands before the King. — *Timbs' "Curiosities of London,"* p. 98.

Christ's Hospital, by ancient custom, possesses the privilege of addressing the Sovereign on the occasion of his or her coming into the City to partake of the hospitality of the Corporation of London. — *Ibid.*

The Dining-Hall, with its lobby and organ-gallery, occupies the entire story, which is 187 feet long, 51 feet wide, and 47 feet high; it is lit by nine large windows, filled with stained glass on the south side; and is, next to Westminster Hall, the noblest room in the metropolis. Here the boys, now about 800 in number, dine; and here are held the "Suppings in Public," to which visitors are admitted by tickets, issued by the Treasurer and by the Governors of Christ's Hospital. The tables are laid with cheese in wooden bowls; beer in wooden piggins, poured from leathern jacks; and bread brought in large baskets. The official company enter; the Lord Mayor, or President, takes his seat in a state chair, made of oak from St. Catherine's Church by the Tower; a hymn is sung, accompanied by the organ; a "Grecian," or head boy, reads the prayers from the pulpit, silence being enforced by three drops of a wooden hammer. After prayer the supper commences, and the visitors walk between the tables. At its close, the "trade-boys" take up the baskets, bowls, jacks, piggins, and candlesticks, and pass in procession, the bowing to the Governors being curiously formal. This spectacle was witnessed by Queen Victoria and Prince Albert in 1845.

Among the more eminent Blue Coat Boys are Joshua Barnes, editor of Anacreon and Euripides; Jeremiah Markland, the eminent critic, particularly in Greek literature; Camden, the antiquary; Bishop Stillingfleet; Samuel Richardson the novelist; Thomas Mitchell, the translator of Aristophanes; Thomas Barnes, many years editor of the London *Times;* Coleridge, Charles Lamb, and Leigh Hunt.

No boy is admitted before he is seven years old, or after he is nine; and no boy can remain in the school after he is fifteen, King's boys and "Grecians" alone excepted. There are about 500 Governors, at the head of whom are the Sovereign and the Prince of Wales. The qualification for a Governor is payment of £500. — *Ibid.*

GENERAL NOTE.

ONE *hears much about the "hideous Blue-Laws of Connecticut," and is accustomed to shudder piously when they are mentioned. There are people in America — and even in England! — who imagine that they were a very monument of malignity, pitilessness, and inhumanity; whereas, in reality they were about the first* SWEEPING DEPARTURE FROM JUDICIAL ATROCITY *which the "civilized" world had seen. This humane and kindly Blue-Law code, of two hundred and forty years ago, stands all by itself, with ages of bloody law on the further side of it, and a century and three-quarters of bloody English law on* THIS *side of it.*

There has never been a time — under the Blue-Laws or any other — when above FOURTEEN *crimes were punishable by death in Connecticut. But in England, within the memory of men who are still hale in body and mind,* TWO HUNDRED AND TWENTY-THREE *crimes were punishable by death!*[1] *These facts are worth knowing — and worth thinking about, too.*

[1] See Dr. J. Hammond Trumbull's "Blue Laws, True and False," p. 11.

FINIS.

AFTERWORD

Everett Emerson

In January 1867 Mark Twain arrived in New York. He had become a professional writer in the West, in Nevada and California, and there he had established a reputation as a humorist, satirist, and critic of the dominant genteel culture. He had ridiculed the clergy; he had attacked romanticism and sentimentality. Applying for a position as editor of a California journal, he had jokingly called himself "The Moral Phenomenon,"[1] but he had remained irreverent. When he moved to the East in 1867 he was very much aware that he needed to clean up his act. In one of his letters for a California newspaper he had commented that a collection of pieces by a Tennessee humorist "will sell well in the West, but the Eastern people will call it coarse and possibly taboo it."[2] When he collected some of his own writings, he and an editor friend blue-penciled them, so that *The Celebrated Jumping Frog of Calaveras County, and Other Sketches* appeared minus allusions to alcohol, sex, gambling, and damnation. A further step toward his gentrification came soon after. While traveling on assignment to Europe and the Holy Land on the *Quaker City*, he met Mary Mason Fairbanks, a fellow passenger — "a most fastidious censor," as he put it.[3] She helped him with the newspaper pieces he was writing and remained both a friend and a critic. But even after Mark Twain had revised his travel pieces, published as *The Innocents Abroad*, the account remained the work of a skeptic and a humorist. His purpose was still to show reality uncolored by pretense, conventionality, and gentility.

His marriage to Olivia Langdon, of Elmira, New York, in 1870 was a marriage to the epitome of the genteel society of the day. Thereafter he was under considerable pressure to write "polite literature." In time he became the father of three daughters, and the pressure increased. After gaining novel-writing experience in composing *The Gilded Age* with a partner, Charles Dudley Warner, Mark Twain wrote his first solo novel, *The Adventures of Tom Sawyer*, published in 1876. Though the book reflected the author's boyhood experiences in a small town on the Mississippi, he himself appears as a writer familiar with the requirements of "every rightly constructed boy's life" (ch. 25) and as an adult familiar with Mont Blanc and wealthy English gentlemen. On the other hand, the reader is never reminded that Mark Twain had known Nevada and California.

After he finished *Tom Sawyer*, Mark Twain began the sequel that would be his masterpiece, *Adventures of Huckleberry Finn*. But many years were to intervene before it was finished, in part because he knew that he ought to be writing something other than a book about the drunkard's son described in *Tom Sawyer* as "dreaded by all the mothers of the town, because he was idle, and lawless, and vulgar and bad" (ch. 6). He wanted to write a book of the sort now expected of him.

While Mark Twain was in England in 1873 he had become interested in the case of Arthur Orton, a Cockney butcher who claimed that he was the heir to the valuable Tichborne estate. The author instructed his secretary "to scrap-book the daily reports of the great trial of the Tichborne Claimant for perjury," as he later recorded in his autobiography.[4] His curiosity about this trial is a reflection of one of Mark Twain's dominant interests, almost an obsessive one, in twins, misplaced children, mistaken identities, impostors, claimants, and the like. The consequences were not only *The Prince and the Pauper* but the Duke and the King in *Huckleberry Finn*, the baby switching of *Pudd'nhead Wilson*, and the doubles and duplicitous identities in *The American Claimant*.

A second influence on *The Prince and the Pauper* was Mark Twain's reading in English history. Sometime around 1874, he read, with remarkable intensity, William E. H. Lecky's *History of European Morals from Augustus to Charlemagne* (1869). By his annotations in the book he indicated his agreement with Lecky's utilitarian notion of moral theory, correcting one formulation of it with a telling deletion: "A desire to obtain happiness and to avoid pain is the only possible motive to action. The reason, and the only reason, why we ~~should~~ perform virtuous actions, or in other words, seek the good of others, is that on the whole such a course will bring us the greatest amount of happiness."[5]

Lecky so aroused his interest in early European history that he read J. A. Froude's *History of England from the Fall of Wolsey to the Defeat of the Spanish Armada* (1870) and the portion of David Hume's *History of England* on Henry VII and Henry VIII. Another work that quite specifically contributed to the novel he was to write was one he read in the summer of 1876: Charlotte M. Yonge's *The Little Duke*, which recounts the adventures of the youthful Duke of Normandy, with emphasis on his moral education. This book gave Mark Twain a model of what would appeal to genteel readers, as well as plot and character ideas.

Mark Twain took very seriously the task of preparing himself to write his historical novel. When he reported his doings to his onetime "fastidious censor," Mary Mason Fairbanks, in February 1877, he knew exactly what he was writing: "A historical tale, of 300 years ago, simply for the love of it — for it will appear without my name — such grave & stately work being considered by the world to be above my proper level. I have been studying for it, off & on, for a year & a half. I swear the Young Girls' Club [a discussion group he organized in 1876] to secresy [*sic*] & read the MS to them, half a dozen chapters at a time, at their meetings. They profess to be very much fascinated with it; so do Livy [his wife] & Susie Warner [a neighbor].[6]

Not only did he read extensively in English history and literature, he made long lists of words and phrases "with the purpose," he later wrote, "of saturating myself with archaic English to a degree which would enable me to do plausible imitations of it in a fairly easy and unlabored way."[7] His lists drew on Samuel Pepys's diary, Shakespeare (especially *Henry IV*), and the novels of Sir Walter Scott. In addition, he observed Scott's advice to a writer not to confine himself to "a particular or limited style."[8]

But as already noted, *The Little Duke* is set in France, and much of what the author had called "studying for" the writing of *The Prince and the Pauper* involved French history and historical literature. In the summer of 1877, in a letter to Mrs. Fairbanks, Mark Twain provided a truly impressive list of his reading, which included works by Victor Hugo, two novels by Alexandre Dumas, Carlyle's "wonderful History of the French Revolution," a biography of Marie Antoinette, a historical novel by Sabine Baring-Gould about the privileges of the French nobles before the Revolution, and some of Hippolyte Taine's *The Ancient Regime*.[9] (One of the British reviewers of Mark Twain's novel was to notice that "the absurd description of the young King's levee" in chapter 14 must have been derived from the author's reading "about the ceremonies of the bedchamber introduced by Louis XIV."[10]) It was from Taine that Mark Twain took the idea that the historian should assist the reader's imagination by providing detailed descriptions of architecture, costumes, the utensils of the hermit's hut, the grounds of the nobleman's estate. (Taine had written that Louis XIV had 383 "officers of the table."[11] In chapter 7 Mark Twain increased the number to 384!) More significantly, like Taine, Mark Twain saw the society he described as polarized into two levels, the privileged and the deprived.

Thus *The Prince and the Pauper* was largely derived not from the author's experience but from his reading, or as he himself had put it, his "studying." In November 1877 he recorded in his notebook the basic plot: "Edward VI & a little pauper exchange places by accident a day or so before Henry III[S] death. The prince wanders in rags & hardships & the pauper suffers the (to him) horrible miseries of princedom, up to the moment of crowning, in Westminster Abbey, when proof is brought & the mistake rectified."[12]

Soon he had written a portion of the novel. Then he stopped, just after the place where Miles Hendon defends Edward before the Guildhall gates; the writing was interrupted by an extended stay in Europe, during which he tried to write a kind of sequel to *The Innocents Abroad*, which he called *A Tramp Abroad*. He took no pleasure from this book, for it proved painfully difficult to complete. In early 1880 he was able to return to his novel, with interest amounting to "intemperance," he told his brother.[13] He wrote his friend

William Dean Howells, "I take so much pleasure in my story that I am loth to hurry, not wanting to get it done." He now had a serious purpose. His idea was "to afford a realizing sense of the exceeding severity of the laws of that day by inflicting some of their penalties upon the king himself & allowing him a chance to see the rest of them applied to others — all of which is to account for certain mildnesses which distinguished Edward VI^{S} reign from those that preceded & followed it." He protested that his "stuff generally gets considerable damning with faint praise" from his wife, but this time, he reported, Livy was "fascinated."[14]

Mark Twain's interest now centered on the painful experiences that the young prince endures as the pauper. To make his account of the "severity of the laws of that day" authentic, he consulted a book by a Hartford neighbor, J. Hammond Trumbull, *The True-Blue Laws of Connecticut and New Haven and the False Blue-Laws* (1876). Trumbull showed that the laws of colonial New England were relatively mild by contrasting them with the English laws of the same period. Since Mark Twain wanted his readers to know that he had based this part of his story on good historical sources, he cited Trumbull and other works freely in his notes, among them Richard Head and Francis Kirkman's *The English Rogue . . . Being a Compleat History of the Most Eminent Cheats of Both Sexes*, published in the 1660s. But the book the author was most indebted to was the aforementioned *History of European Morals from Augustus to Charlemagne*, one of his all-time favorites. From his reading in Lecky, he came to recognize that human progress depends on education and experience, a recognition that shapes *The Prince and the Pauper*, in which both factors loom large. Thus Edward pledges that the Christ's Hospital boys "shall not have bread and shelter only, but also teachings out of books; . . . for learning softeneth the heart and breedeth gentleness and charity" (52–53). The prince learns compassion through his adventures.

In 1880, Mark Twain received several letters that bolstered his feeling that he was now using his talents wisely. A Hartford Minister, Edwin P. Parker, asked that he do himself "vast honor" and give his friends "vast pleasure" by writing a book with "a sober character."[15] Mary Mason Fairbanks argued that he should write "another book, . . . in an entirely different style. . . . The time has come for your *best book*. . . . your best contribution to American literature."[16]

By December 1880 the book was ready for his friend and critic Howells to read. Howells gave the manuscript a close perusal, pronounced it "good," but noted that it was "long-winded in some places."[17] The author accepted many of Howells' criticisms but expanded sections of the book as well, adding "over 130 new pages of MS to the prince's adventures in rural districts."[18] He also sought out the judgments of Howells' children and those of his minister friends Joseph Twichell and Edwin Parker. When he was finally through with his task, he declared, "I like this book better than *Tom Sawyer*—because I haven't put any fun in it. I think *that* is why I like it better. You know a body always enjoys seeing himself attempting something out of his line."[19] He wrote his sister that his wife was eager to have the book "elegantly gotten up, even if the elegance of it eats up the publisher's profits and mine too."[20]

Because it was indeed "out of his line" and because his publisher, Elisha Bliss of the American Publishing Company, had died in 1880, Mark Twain arranged for James R. Osgood of Boston to publish his book; Osgood had considerable prestige. Bliss's company had published *The Innocents Abroad*, *Roughing It*, *The Gilded Age*, *Tom Sawyer*, and *A Tramp Abroad*, all by subscription. Under this arrangement, representatives of the publisher would visit small towns and walk door to door with samples of the latest Mark Twain book, to take orders before publication. The books were not sold in stores. A customer could even select the binding, which might be of expensive leather if the customer was well-to-do. As a result of this sales system, which had proved profitable to him, Mark Twain was widely known among ordinary people; his books were liked by people who were not bookish.

Osgood, however, had no experience as a subscription publisher and was ill equipped for his task. Moreover, Mark Twain changed the usual terms of subscription publishing, stipulating that he would hold the copyright, not Osgood, and contracting to pay for all of the publishing costs, with Osgood receiving a royalty of 7.5 percent on books sold. Like other subscription books, *The Prince and the Pauper* would be illustrated; indeed, illustrations had been an important feature of all of Mark Twain's books. Now he could control this aspect of the publication process. He specified that "the artist always picture the Prince and Tom Canty as lads of 13 or 14 years."[21] (The historical prince was less than ten years old.) Probably he intended his book to appeal especially to teenagers.

The illustrators were John Harley and Frank T. Merrill. Both were competent, and the illustrations are a most appealing feature. The work of the two artists can be distinguished: Harley's figures are much sturdier than Merrill's—compare Merrill's chapter 3 illustration captioned "Doff Thy Rags, and Don These Splendors" (43) with Harley's chapter 13 illustration captioned "Hendon Followed After Him" (156). Mark Twain especially admired Merrill's work: "Merrill probably thinks he *originated* his exquisite boys himself, but I was ahead of him there!—in these pictures they look and dress exactly as I used to see them in my mind two years ago. It is a vast pleasure to see them cast in the flesh, so to speak—they were of but perishable dream-stuff before."[22] (Merrill's earlier illustrations for Louisa May Alcott's *Little Women* are famous.)

On the title page, *The Prince and the Pauper* is identified as "A Tale for Young People of All Ages." The dedication reads, "To those good-mannered and agreeable children, Susie and Clara Clemens, this book is affectionately inscribed by their father." At the time of publication, Olivia Susan, known as Susy (at first "Susie"), was nine and Clara was seven. Susy later recorded, "I have wanted papa to write a book that would reveal his kind sympathetic nature, and the 'Prince and Pauper' partly does it. The book is full of lovely charming ideas, and oh the language! it is perfect, I think."[23] (Mark Twain supposed for a time that his story might well be published in the *St. Nicholas Magazine for Boys and Girls*, though he finally decided that its appearance there would hurt book sales.)

In 1882 Mrs. Fairbanks recalled how the author had read his story aloud, "Livy on one side of the fire, and those honest critics Susie and Clara perched on your arm-chair."[24] And Mark Twain had asked his friends to read his story to their families. Although he hoped to appeal to the growing juvenile market, he wanted to address his story to both children and adults, for whom he provided portrayals of proper parents, specifically Miles Hendon, in his fathering of the prince (which resembles Jim's fathering of Huck), and Tom Canty's mother.

In the late nineteenth century, many genteel writers were unhappily aware that the most readily available children's books were the popular Beadle's dime novels that told of rebels against propriety, law, and authority, such as Deadwood Dick, the protagonist of thirty-three episodes by E. L. Wheeler. In these books, fathers and sons were presented as natural enemies, and respectable family life was wholly absent. One commentator, Horace Scudder, complained, "The most popular books for the young are those which represent boys and girls as seeking their fortune, working out their own schemes," apart from their elders.[25] It was in this context that Mark Twain wrote *The Prince and the Pauper*, in which, significantly, the prince remembers "a long succession of loving passages between his father and himself, and [dwells] fondly on them." (Most readers would not naturally assume that Henry VIII was a kindly father.)

With a setting in the mid-sixteenth century, the time of Edward VI, so remote from Clemens' own experience that he documents it with footnotes citing authorities, the book provides no sense of being present at the events it describes. The writer whose contribution to American literature is the colloquial style here produces dialogue that is a labored attempt at Elizabethan English. "Searched you well—but it boots not to ask that. It doth seem passing strange" (376). A few metaphors reveal the writer's background. If the guardians of Tom Canty, the pauper who by mischance becomes the prince, "felt much as if they were piloting a great ship through a dangerous channel," for Tom himself, time "wore on pleasantly, and likewise smoothly, on the whole. Snags and sandbars grew less and less frequent" (78). With *The Prince and the Pauper* Mark Twain was out of his element, and his old friend Joe Goodman, from his days in Virginia City, Nevada, told him so: "But what could have sent you groping among the driftwood of the Deluge for a topic when you could have been so much more at home in the wash of today?"[26]

When the novel appeared, the *Hartford Courant* congratulated the author for "writing a book which has other and higher merits than can possibly belong to the most artistic expression of mere humor."[27] His neighbor Harriet Beecher Stowe told him it was "the best book for young folks that was ever written."[28] Kenneth R. Andrews, in his study of Mark Twain's cultural milieu, says *The Prince and the Pauper* was "precisely gauged to the taste of Hartford."[29] Only Joe Goodman is recorded as thinking the book a mistake.

Just below the surface one can detect much that reflects the author's interests and concerns. The courtly ceremonies of Europe had attracted him as early as his first visit, when he saw Napoleon III and met the czar of Russia, as well as later when he covered the tour of the shah of Persia. Increased financial pressures on Clemens are reflected in Tom's worried comments in chapter 14 on royal expenditures: "We be going to the dogs, 'tis plain. 'Tis meet and necessary that we take a smaller house and set the servants at large" (167). Like Tom Sawyer (and the author), Tom Canty yearns for excitement and is bored by routine, and both boys are basically good-hearted, innocent. Their goodness is underscored by the cruelty of the society in which they live, a concern soon to become a major preoccupation for their creator.

Mark Twain had always been sensitive to suffering, as his writings about the Chinese in California show. Now his reading in historical works had helped him to see, as he told Howells in March 1880, "the exceeding severity of the laws of that day." Moreover, he was disenchanted with monarchy if not with its trappings. What the prince learns of suffering and oppression makes him a merciful king when the two boys change places a second time. And the education the prince receives is more than an indictment of a system of government or a historical epoch. Unlike Charlotte Yonge's *The Little Duke*, *The Prince and the Pauper* depicts man's cruelty as part of his nature. Though a book for children, it is the first of Mark Twain's works to castigate the damned human race, as he was to call it. Like Huck Finn, the prince learns the facts of life through his travels, and like Huck, he is accompanied by an older man. (Later Mark Twain would return to precisely this theme to tell how another ruler's eyes were opened by his experiences while traveling incognito with the Connecticut Yankee.) The tone, however, is far more optimistic than that of the later books, though the reader may see that the lessons the prince learns will have little effect, because the real Edward VI reigned only briefly and died at the age of fifteen.

Though *The Prince and the Pauper* marks a turn in Mark Twain's literary career, it looks backward too. Like Mark Twain in *The Innocents Abroad* and the tenderfoot narrator in *Roughing It* (the book about the author's Western experiences), both boys at the beginning of their adventures have unrealistic anticipations that distort their vision; Edward, of course, expects to be treated like a king, and Tom has listened "to good Father Andrew's charming old tales and legends about giants and faeries, dwarfs and genii, and enchanted castles, and gorgeous kings and princes" (29–30). What happens to the two boys, especially the prince, serves as an educational corrective.

Unlike most of Mark Twain's longer works, whether fiction or nonfiction, *The Prince and the Pauper* is well plotted. The experiences of the prince and the pauper after they exchange roles are neatly parallel: each boy finds that his "father" believes him to be mad, each is befriended by his "sister," and each wakes from sleep thinking that his trials have been just a bad dream. Though the reader assumes that the two boys will revert to their original roles, suspense arises in the uncertainty over exactly how that development is to occur. Only at the last possible moment do the boys resume their proper places.

The most important theme of *The Prince and the Pauper* is one the author would return to often: the mystery of identity. The switching of roles that forms the basis of the plot permitted Clemens to demonstrate what was becoming one of his pet ideas, later set forth simply: "Training is everything." The differences between the prince and the pauper are only skin-deep, for the boys are still young. The pauper quickly learns to play the role of prince, as a result of the training he receives. Later Clemens would perform a more radical experiment by switching a "black" slave and the son of a white aristocrat while they are still babies.

More significantly, the author's own life experiences brought him to his theme. After all, he had created a pen personality: as Bret Harte had reminded his readers, "Mark Twain" was "a very eccentric creation of Samuel Clemens."[30] After Clemens met Olivia Langdon, he was eager to adopt genteel manners, and he can be said to have met the challenge with great success.

But there is evidence that his transformation came at a price. In his last years Mark Twain began an unfinished novel called *Indiantown* that suggested his discomfort with his metamorphosis from Western humorist to Eastern gentleman. Here he describes one David Gridley, whose wife, Susan, has converted him into an "elaborate sham," a replacement for the man she married: "As far as his outside was concerned she made a masterwork of it that would have deceived the elect." "The real David, the inside David, the hidden David, was of an incurably low tone, and wedded to low ideals; the outside David, Susan Gridley's David, the sham David, was of a lofty tone, with ideals which the angels in heaven might envy." Significantly, David's wife "edited David's letters for him, but not by request."[31]

Perhaps the most subtle aspect of the treatment of identity in *The Prince and the Pauper* is the demand of three central characters to be recognized for who they are. The prince and the pauper both plead for recognition, and so does Miles Hendon. The reader can see in their protests a reflection of the author's desire to have his true literary identity recognized. The problem was that "Mark Twain" remained a mystery, even to the writer himself. On his deathbed, we are told by his biographer Albert Bigelow Paine—who was there—he talked about "Dual Personality, and discussed various instances that flitted through his mind—Jekyll and Hyde phases in literature and in fact."[32] *The Prince and the Pauper* is a significant demonstration of the author's vital theme.

In what ways *The Prince and the Pauper* is a book designed to satisfy the genteel readers of Hartford and elsewhere is not hard to understand. It is a historical tale laden with facts about the past and set in England. Thus it appealed to those burdened by a need to learn from a novel as well as to be entertained, especially those Americans whose model was English gentility. Clemens himself had become an ardent admirer of England, though before he finished the book his admiration had cooled. He was well acquainted with England and had broken bread with many of its aristocrats. Having dredged up much historical material, he was able to show off, like Tom Sawyer, even if his learning was precariously balanced. His notebook contains a list of over seventy people to whom he proudly sent copies of his book, among them four of the great New England writers: Holmes, Emerson, Whittier, and Longfellow. It must have been particularly gratifying to Clemens to address them in this way after the humiliation of the Whittier birthday dinner of 1877. Copies were also sent to the Scottish clergyman-poet George MacDonald, to Charlotte Yonge, and to the writer Rose Terry Cooke. Specially printed and bound copies went to the good-mannered Susy and Clara. He had done something special — something out of his line — and he wanted the world to know.

Because of Clemens' ambitions for the book, encouraged by his wife, production costs were high. These expenses were borne by the author, who had thought he could increase his profits through such a financial arrangement. But because his readers did not expect him to write such a work and because of Osgood's lack of publishing knowledge, *The Prince and the Pauper* was a financial failure. Over 21,000 copies of the book were sold between the publication date, December 12, 1991, and March 1, 1882, but sales fell dramatically thereafter. Some 5,000 copies still unsold by February 1884 were transferred to Charles L. Webster and Company, the publishing house created by Clemens with his sister's son-in-law as chief executive officer. As Mark Twain later explained,

> Osgood was one of the dearest and sweetest and loveliest human beings to be found on the planet anywhere, but he knew nothing about subscription publishing and he made a mighty botch of it. He was a sociable creature and we played much billiards and daily and nightly had a good time. And in the meantime his clerks ran our business for us and I think that neither of us inquired into their methods or knew what they were doing.[33]

In his autobiography, Mark Twain reported that he earned a mere $17,000 total on the American edition of the book.

Among the earliest reviews to appear was that of Hjalmar Hjorth Boyesen, a professor of German and a novelist. In the December 1881 *Atlantic Monthly*, in a review entitled "Mark Twain's New Departure," he called *The Prince and the Pauper* "a tale ingenious in conception, pure and humane in purpose, artistic in method, and, with barely a flaw, refined in execution."[34] Mark

Twain must have been thoroughly gratified. The reviewer for *Harper's* likewise admired the book, which was found to be "rich in historical facts and teachings."[35] His friend Howells had written an anonymous review that appeared in the *New York Daily Tribune* in October 1881, well before publication, in an attempt to prepare the public for a "serious" Mark Twain. Howells pointed to "the strain of deep earnestness underlying [the] humor," words that must have delighted the author's wife, and ended by saying, "It is a beautiful story: airy and lawless as an Arab tale in conception; but so solidly good and wholesome in effect that one wishes it might have happened."[36] The Southern writer Joel Chandler Harris, in a review in the *Atlanta Constitution*, saw the book as a "wide departure" for Mark Twain, whom he called "a true literary artist."[37] The *Century* was less admiring; in its view the author's departure from his accustomed manner was not "a conspicuous success."[38]

Across the water, the London *Times* described the story as "capitally told, in an easy and picturesque style."[39] The *Graphic* of London also praised the book: "its humor is delicate; its fun joyously real; and its pathos tender and deep."[40] But several other publications were far more critical. The *British Quarterly Review* reported that Mark Twain "has written a story which gives a vivid idea of a historical period, but which outrages history at every point in the most daring and barefaced manner."[41] The *Athenaeum* called the book "a heavy disappointment," at "some four hundred pages of careful tediousness."[42] The *Academy* was even more severe: E. Purcell, the reviewer who noticed the incongruities in "the young King's levee," condemned the novel as "a ponderous fantasia of English history," "a prolix work singularly deficient in literary merit."[43] The book continued, nonetheless, to have English enthusiasts. One of them, the Reverend H. R. Haweis, commented in an 1881 London lecture, "What can be more unlike any of his previous works than 'The Prince and the Pauper,' Mark Twain's 1881 Christmas book? There the fun lies in the fancy that Edward VI., in a freak just before he ascended the throne, changed clothes with a romantic beggar who resembled him closely."[44]

In time *The Prince and the Pauper* established itself; it is now among the most widely read of Mark Twain's books in the United States. Moreover, it is popular not only in English-speaking parts of the world but beyond: it has been translated into Danish, Finnish, Italian (over fifty editions), and German (over forty editions), and has had multiple editions in French, Spanish, Polish, Russian, and Japanese — a total of 550 foreign editions through 1976. In recent years the book has been much reprinted, often in editions clearly designed for children. But as the author insisted, it is a book "for young people of all ages."

Why is *The Prince and the Pauper* still widely read? The answer is not far to seek. It is a highly entertaining and attractive story, with a strong sense of moral principles at its center. Like *Tom Sawyer* and *Huckleberry Finn*, it focuses on children, but this time Mark Twain takes the reader not to his own middle America but across the water and into the past. Thus the novel shows something of the richness of the author's interests. If it lacks the humor of much of Mark Twain's writing, it compensates by providing an excellent plot and lively suspense. And it introduces readers to one of Mark Twain's major interests, still of great appeal: the nature of identity and its relationship to environment. *The Prince and the Pauper* remains a classic.

NOTES

1. Mark Twain, *Early Tales and Sketches, Volume 2* (1864–1865), ed. Edgar Branch and Robert H. Hirst (Berkeley: University of California Press, 1981), pp. 54–55.

2. *Mark Twain's Travels with Mr. Brown*, ed. Franklin Walker and G. Ezra Dane (New York: Knopf, 1940), p. 221.

3. Letter of November 22, 1867; *Mark Twain's Letters, Volume 2 (1867–1868)*, ed. Harriet Smith and Richard Bucci (Berkeley: University of California Press, 1990), p. 108.

4. *Mark Twain's Autobiography*, ed. Albert Bigelow Paine (New York: Harper, 1924), 1:139.

5. Quoted in Mary Boewe, "Twain on Lecky: Some Marginalia at Quarry Farm," *Mark Twain Society Bulletin* 8 (January 1985): 3.

6. *Mark Twain to Mrs. Fairbanks*, ed. Dixon Wecter (San Marino, Calif.: Huntington Library, 1949), p. 218.

7. Autobiographical dictation published in *Mark Twain in Eruption*, ed. Bernard DeVoto (New York: Harper, 1940), p. 206.

8. Scott, introduction to *Ivanhoe*, The Waverly Novels, Vol. 9, autograph edition (Edinburgh: Black, n.d.), pp.1–2.

9. *Mark Twain to Mrs. Fairbanks*, pp. 207–8.

10. E. Purcell's review of *The Prince and the Pauper*, *Academy* (London) 20 (December 24, 1881): 469.

11. Hippolyte Taine, *The Ancient Regime*, trans. John Durand (New York: Henry Holt, 1876), p. 95.

12. *Mark Twain's Notebooks and Journals, Volume 2 (1877–1883)*, ed. Frederick Anderson, Lin Salamo, and Bernard L. Stein (Berkeley: University of California Press, 1975), p. 49.

13. Letter dated February 20, 1880; Vassar College; quoted in Everett Emerson, *The Authentic Mark Twain: A Literary Biography of Samuel L. Clemens* (Philadelphia: University of Pennsylvania Press, 1984), p. 107.

14. Letter of March 11, 1880; *Mark Twain–Howells Letters*, ed. Henry Nash Smith and William M. Gibson (Cambridge: Harvard University Press, 1960), 1:291–92.

15. Letter of December 22, 1880; Mark Twain Papers (hereafter cited as MTP), University of California, Berkeley.

16. Letter of July 26, 1880, MTP.

17. Letter of December 15, 1880; *Mark Twain–Howells Letters*, 1:339.

18. Letter dated January 21, 1881; copy at MTP; quoted in Emerson, *The Authentic Mark Twain*, p. 108.

19. Letter of January 31, 1881; published in *Mark Twain the Letter Writer*, ed. Cyril Clemens (Boston: Meador, 1932), p. 37.

20. Quoted in Albert Bigelow Paine, *Mark Twain: A Biography* (New York: Harper, 1912), 2:696.

21. Letter of March 9, 1881; quoted in *The Prince and the Pauper*, ed. Victor Fischer and Lin Salamo (Berkeley: University of California Press, 1979), p. 11.

22. Letter of August 14, 1881; *Mark Twain's Letters to His Publishers*, ed. Hamlin Hill (Berkeley: University of California Press, 1967), p. 140.

23. Susy Clemens, *Papa: An Intimate Biography of Mark Twain*, ed. Charles Neider (Garden City, N.Y.: Doubleday, 1985), p. 107.

24. Letter of January 4, 1882; quoted in *Mark Twain to Mrs. Fairbanks*, p. 245, n. 1.

25. Scudder, *Childhood in Literature and Art*; quoted in Albert E. Stone, *The Innocent Eye: Childhood in Mark Twain's Imagination* (New Haven: Yale University Press, 1961), p. 107.

26. Letter of October 24, 1881, MTP.

27. *Courant*, December 28, 1881.

28. Recorded by the author; *Mark Twain's Notebooks and Journals, Volume 3 (1883–1891)*, ed. Robert Pack Browning et al. (Berkeley: University of California Press, 1979), p. 287.

29. *Nook Farm: Mark Twain's Hartford Circle* (Cambridge: Harvard University Press, 1950), p. 194.

30. *Overland Monthly*, January 1870.

31. *Mark Twain's Which Was the Dream? and Other Symbolic Writings of the Later Years*, ed. John S. Tuckey (Berkeley: University of California Press, 1968), pp. 166–70.

32. Paine, *Mark Twain*, 3:1575.

33. Autobiographical dictation, *Mark Twain in Eruption*, p. 157.

34. *Atlantic Monthly* 48 (December 1881): 843–45.

35. *Harper's* 64 (March 1882): 635.

36. *New York Daily Tribune*, October 25, 1881, p. 6.

37. *Atlantic Constitution*, December 25, 1881.

38. *Century* 23 (March 1882): 783–84.

39. *Times*, December 20, 1881, p. 3.

40. *Graphic* 25 (March 25, 1881): 306.

41. *British Quarterly Review* 75 (January 1882): 118.

42. *Athenaeum* December 24, 1881, 849.

43. Purcell, *Academy* 20:469.

44. "Mark Twain," in *American Humorists: Lectures Delivered at the Royal Institution* (London: Chatto and Windus, 1883), p. 64.

FOR FURTHER READING

Everett Emerson

The doubleness that is suggested even by the author's having two names is central to two biographies of the man and the writer. Justin Kaplan called his book *Mr. Clemens and Mark Twain* (New York: Simon and Schuster, 1966) to suggest the extent to which Samuel Clemens/Mark Twain was a double creature: "The Hartford literary gentleman lived inside the sagebrush bohemian." In my literary biography, *The Authentic Mark Twain* (Philadelphia: University of Pennsylvania Press, 1984), I explore the difficulties that the author and man faced as he tried to do justice to both his sense of his inner self and his social ambitions. The theme is taken up also in Susan Gillman's *Dark Twins: Imposture and Identity in Mark Twain's America* (Chicago: University of Chicago Press, 1989), which focuses especially on *Pudd'nhead Wilson* and tales of sexual identity.

Mark Twain's relationship to Hartford is described in Kenneth R. Andrews, *Nook Farm: Mark Twain's Hartford Circle* (Cambridge: Harvard University Press, 1950). His relationship to England as it affected *The Prince and the Pauper* is discussed by Howard G. Baetzhold, *Mark Twain and John Bull: The British Connection* (Bloomington: Indiana University Press, 1970). *The Prince and the Pauper* is seen in the context of the development of Mark Twain's thinking in Sherwood Cummings, *Mark Twain and Science: Adventures of a Mind* (Baton Rouge: Louisiana State University Press, 1988). A full-dress scholarly edition of *The Prince and the Pauper* was published by the University of California Press in 1979; it was edited by Victor Fischer and Lin Salamo.

ILLUSTRATORS AND ILLUSTRATIONS IN MARK TWAIN'S FIRST AMERICAN EDITIONS

Beverly R. David & Ray Sapirstein

From the "gorgeous gold frog" stamped into the cover of *The Celebrated Jumping Frog of Calaveras County* in 1867 to the comet-riding captain on the frontispiece of *Extract from Captain Stormfield's Visit to Heaven* in 1909, illustrators and illustrations were an integral part of Mark Twain's first editions.

Twain marketed most of his major works by subscription, and illustration functioned as an important sales tool. Subscription books were packed with pictures of every type and size and were bound in brassy gold-stamped covers. The books were sold by agents who flipped through a prospectus filled with lively illustrations, selected text, and binding samples. Illustrations quickly conveyed a sense of the story, condensing the proverbial "thousand words" and outlining the scope and tone of the work, making an impression on the potential purchaser even before the full text had been printed. Book canvassers were rewarded with up to 50 percent of the selling price, which started at $3.50 and ranged as high as $7.00 for more ornate bindings. The books themselves were seldom produced until a substantial number of customers had placed orders. To justify the relatively high price and to reassure buyers that they were getting their money's worth, books published by subscription had to offer sensational volume and apparent substance. As Frank Bliss of the American Publishing Company observed, these consumers "would not pay for blank paper and wide margins. They wanted everything filled up with type or pictures." While authors of trade books generally tolerated lighter sales, gratified by attracting a "better class of readers," as Hamlin Hill put it, authors of subscription books sacrificed literary respectability for popular appeal and considerable profit.[1]

The humorist George Ade remembered Twain's books vividly, offering us a child's-eye view of the nineteenth-century subscription book market.

> Just when front-room literature seemed at its lowest ebb, so far as the American boy was concerned, along came Mark Twain. His books looked at a distance, just like the other distended, diluted, and altogether tasteless volumes that had been used for several decades to balance the ends of the center table . . . so thick and heavy and emblazoned with gold that [they] could keep company with the bulky and high-priced Bible. . . . The publisher knew his public, so he gave a pound of book for every fifty cents, and crowded in plenty of wood-cuts and stamped the outside with golden bouquets and put in a steel engraving of the author, with a tissue paper veil over it, and "sicked" his multitude of broken-down clergymen, maiden ladies, grass widows, and college students on the great American public.
>
> Can you see the boy, Sunday morning prisoner, approach the book with a dull sense of foreboding, expecting a dose of Tupper's *Proverbial Philosophy*? Can you see him a few minutes later when he finds himself linked arm-in-arm with Mulberry Sellers or Buck Fanshaw or the convulsing idiot who wanted to know if Christopher Columbus was sure-enough dead? No wonder he curled up on the hair-cloth sofa and hugged the thing to his bosom and lost all interest in Sunday school. *Innocents Abroad* was the most enthralling book ever printed until *Roughing It* appeared. Then along came *The Gilded Age*, *Life on the Mississippi*, and *Tom Sawyer*. . . . While waiting for a new one we read the old ones all over again.[2]

Publishers, editors, and Twain himself spent a good deal of time on design—choosing the most talented artists, directing their interpretations of text, selecting from the final prints, and at times removing material they deemed unfit for illustration.[3]

With the exception of *Following the Equator* (1897), books released in the twilight of Twain's career were not sold by subscription. Twain's later books, published for the trade market by Harper and Brothers, seldom contained more than a frontispiece and a dozen or so tasteful illustrations, rather than the hundreds of illustrations per volume that subscription publishing demanded. Illustration, however, remained a major component of Twain's later work in two important cases: *Extracts from Adam's Diary*, illustrated by Fred

Strothmann in 1904, and *Eve's Diary*, illustrated by Lester Ralph in 1906.

The stories behind the illustrators and illustrations of Mark Twain's first editions abound in back-room intrigue. The besotted or negligent lapses of some of the artists and the procrastinations of the engravers are legendary. The consequent production delays, mistimed releases, and copyright infringements all implied a lack of competent supervision that frequently infuriated Twain and ultimately encouraged him to launch his own publishing company.

In many cases, Twain took illustrations into account as he wrote and edited his text, using them as counterpoint and accompaniment to his words, often allowing them to inform his general narrative strategy and to influence the amount of detail he felt necessary to include in his written descriptions. In the most artful and carefully considered illustrated works, an analysis of the relationships between author and illustrator and between text and pictures illuminates key dimensions of Twain's writings and the responses they have elicited from readers. Examinations of even the most straightforward examples of decorative imagery yield insights into the publishing history of Twain's books and his attitudes toward the production process.

The original illustrations in Twain's works have often been replaced in the twentieth century by subsequent visual interpretations. But while Norman Rockwell's well-known nostalgic renderings of *Tom Sawyer* and *Huckleberry Finn* may tell us much about 1930s sensibilities, we would do well to reacquaint ourselves with the first American editions and the artwork they contained if we want to understand the books Twain wrote and the world they affected.

Illustrated books, like the illustrated weekly magazines that first appeared in the 1860s, were a significant source of visual images entering nineteenth-century homes. Because of their widespread popularity and the relative paucity of other sources of visual information, Twain's books helped to define America's perceptions of remote people, exotic scenes, and historic events. In addition to being an essential element of Mark Twain's body of work, illustrations are a documentary source in their own right, a window into Twain's world and our own.

NOTES

1. For background on subscription book publishing, see Hamlin Hill, *Mark Twain and Elisha Bliss* (Columbia: University of Missouri Press, 1964), chapter 1. See also R. Kent Rasmussen, "Subscription-book publishing" entry, *Mark Twain A to Z: The Essential Reference to His Life and Writings* (New York: Facts on File, 1995), p. 448.

2. George Ade, "Mark Twain and the Old-Time Subscription Book," *Review of Reviews* 61 (June 10, 1910): 703–4; reprinted in Frederick Anderson, ed., *Mark Twain: The Critical Heritage* (London: Routledge and Kegan Paul, 1971), pp. 337–39.

3. Beverly R. David, *Mark Twain and His Illustrators, Volume 1 (1869–1875)* (Troy, N.Y.: Whitston Publishing Company, 1986), discusses in detail Twain's involvement in the production of his early books.

READING THE ILLUSTRATIONS IN *THE PRINCE AND THE PAUPER*

Beverly R. David & Ray Sapirstein

The Prince and the Pauper (1881) found Mark Twain with a new publisher, James R. Osgood, who headed the trade publishing house that bore his name. Twain, who did not want to give up the hefty profits he had come to expect from the subscription market, persuaded Osgood to publish *The Prince and the Pauper* by subscription. This arrangement yielded two primary benefits: Osgood and Twain could discuss business while playing billiards, and Andrew Varick Stout Anthony could relieve Twain of the time-consuming tasks of hiring and supervising artists and overseeing the details of production. Anthony (1835–1906) was an extremely capable and conscientious supervisor. He was an engraver, and had been an illustrator himself; his credits included work on Longfellow's "The Skeleton in Armor." As Osgood's longtime design director, Anthony was the creative force behind Twain's most elegant and ornate books, *The Prince and the Pauper*, *The Stolen White Elephant*, and *Life on the Mississippi*.

Anthony hired three illustrators — Frank T. Merrill, John Harley, and L. S. Ipsen — whom Twain touted as "high priced artists and engravers."[1] Their fees naturally increased production costs, which Twain bore himself; he estimated that manufacturing *The Prince and the Pauper* cost seventy cents per copy, well above average for his previous subscription books.[2]

Frank Merrill was the principal illustrator, despite the fact that at thirty-two he had not yet amassed an extensive list of credits. He had, however, illustrated Louisa May Alcott's *Little Women* and *Moods*, published in 1868 and 1870 respectively. He was responsible for drawing the members of the court and their attendants in historically appropriate costumes and settings. Much of Merrill's work is signed. John Harley, a Canadian who had done cuts for James Fenimore Cooper's *The Prairie* for Appleton, and for Harriet Beecher Stowe's *Oldtime Fireside Stories*, published by Osgood, illustrated most of the scenes of the lower classes, including the views of Offal Court and Cheapside. Harley's work is signed with an "H." L. S. Ipsen, a relatively unknown artist who a year later would collaborate with Anthony on an exquisite edition of Sir Walter Scott's *The Lady of the Lake*, was assigned to design the chapter headings, the "half titles" itemized in the list of illustrations. Each chapter title features calligraphic lettering emblazoned across elegant scenes, emblems, or still-life images related to the action, bordered by elegantly textured, three-dimensional decorative frames. More than the drawings, which frequently venture into nineteenth-century melodrama and bathos, the chapter titles give the book an aura of fine craftsmanship, literary substance, and historical resonance.

Ipsen designed a suitably regal cover that displayed all the accoutrements of the English sovereign: a golden crown, shield, scepter, ball, and bludgeon. The title and Twain's name were spelled out in ornately seriffed gilt letters. Twain decided to include several frontispieces in the first edition. The Latimer Letter, in both original and transcription, was copied directly from James' *Facsimiles of National Manuscripts* and placed before the title page. The great seal of Henry VIII was reproduced from an engraving in Craik's *Pictorial History of England.*[3] A perfect signature by the monarch was reproduced below the seal, his "H" formed without lifting the pen from the writing surface. Twain cautioned that the great seal "wasn't to be engraved — old Brer Osgood forgot about that, I reckon. I'm afraid to put it in."[4] He was apparently worried about offending his large British audience and blunting the impact of the coronation scene, in which the seal figures as the critical instrument that differentiates the two boys. The author's fears were quieted, and the seal and autograph appeared opposite the title page.

The rendering of the figures of the two young heroes was divided between Merrill and Harley, with resulting discrepancies in characterization. Admitting that he had not stipulated an age for the pair in the text, Twain finally wrote the publishers asking that "the artist always picture the Prince & Tom Canty as lads of 13 or 14 years old."[5] However, when Anthony saw the first drawings submitted by Harley, he suggested that his Tom looked too young, more like a boy of ten or eleven. The next prints of the royal household, by

Merrill, found the prince a lanky adolescent of about fourteen. This inconsistency continued throughout the book, probably because Anthony was more concerned with the quality of the artwork than with the details of characterization. In the end, Twain focused his praise upon Merrill's delicate figures.

> Merrill probably thinks he *originated* his exquisite boys himself, but I was ahead of him there! — in these pictures they look and dress exactly as I used to see them in my mind two years ago. It is a vast pleasure to see them cast in the flesh. . . . They were of but perishable dream-stuff before.[6]

He asked the publishers to "glorify the illustrations" in the salesmen's prospectus and "call attention to the historical accuracy of the costumes." He added, "The more I examine the pictures, the more I am enchanted with them."[7]

Historical accuracy was a crucial issue for Twain. He had given the publishers several old English books for the artists' reference, including Timbs' *Curiosities of London* and the *Pictorial History of England.* The burning of Anne Askew (31) featured background buildings from "The Burning Place in Smithfield" in the *Pictorial History.* The view of Offal Court (28) and the London scene that appears in Ipsen's "The Birth of the Prince and the Pauper" (21) were undoubtedly based on pictures from the same volume. The costume of the Christ's Hospital orphans was copied from *Curiosities of London*, and in the notes appended to *The Prince and the Pauper* Twain added a section on this costume taken from Timbs. The ermine-robed Henry VIII and the veiled and swaddled ladies of the court could have come straight out of the famous Holbein portraits commonly engraved in history books of the period.

The completed 412-page volume contained 192 illustrations, a rather meager ratio by subscription publishing standards. But Twain wrote that the "dainty and rich" illustrations "clear surpass my highest expectations."[8] He had a small private edition of the book printed, and distributed copies as gifts to family and friends.[9]

The Prince and the Pauper was a departure for Twain in several important aspects. It was a fresh start with a new publisher, his first major book after he left Elisha Bliss's American Publishing Company, which had brought out all his books since *Innocents Abroad* in 1869. While Twain sought to retain the profits of subscription publishing, Osgood's Boston base and his reputation as one of the foremost American literary publishers lent Twain a more cultured image than the one he had cultivated as a writer of comic sketches. He intended *The Prince and the Pauper* as a novel of lasting social and historical merit that would earn him recognition as an author firmly moored in the belles lettres tradition. The illustrations, as well as the text, demonstrate the breadth of Twain's new bid for literary respectability.

When Twain described his illustrators as "high-priced," he meant to imply as well that they were high-toned. It was not lost on him that these artists had illustrated works by Longfellow, Stowe, Cooper, Scott, and Alcott, writers who occupied secure seats in the literary pantheon to which he himself now aspired. Along with Osgood's publishing credits and author list, the illustrators were part of the literary status the firm offered Twain. Rather than the cartoonish visual high jinks and quickly rendered images produced for him during his decade-long tenure at the American Publishing Company, exactingly crafted and soberly judicious images now accompanied Twain's words. Elisha Bliss may have boasted of outstanding artwork as part of his promotional campaigns for Twain's earlier books, but no brassy sales pitch was required to convey the quality of the illustrations here. Framing Twain's text, the elegant images by eminent artists spoke for themselves—and required the reader to take Twain seriously as a writer of substance.

NOTES

1. Mark Twain to Pamela Moffett, March 16, 1881, in *Mark Twain, Business Man*, Samuel Charles Webster, ed. (Boston: Little, Brown, 1946), p. 150.

2. *Introduction to The Prince and the Pauper*, ed. Victor Fischer and Lin Salamo, The Works of Mark Twain (Berkeley: University of California Press, 1979), p. 10, n. 34.

3. The motto of the Order of the Garter was inscribed on the seal: *Honi Soit qui mal y pense* (Evil to him who evil thinks). Twain and Dan Beard later parodied the motto in an illustration for *A Connecticut Yankee*, "Decorations of the Sixth Century Aristocracy," that listed the abuses of the English nobility.

4. Mark Twain to Benjamin Ticknor, August 14, 1881, in *Mark Twain's Letters to His Publishers, 1867–1894*, Hamlin Hill, ed. (Berkeley: University of California Press, 1967), p.140.

5. Cited in the introduction to *The Prince and the Pauper*, p. 11.

6. Ibid.

7. S. L. Clemens to Benjamin Ticknor, August 14, 1881, in *Mark Twain's Letters to His Publishers*, p.139.

8. S. L. Clemens to Benjamin Ticknor, August 1, 1881 in *Mark Twain's Letters to His Publishers*, p.138.

9. Merle Johnson, *A Bibliography of the Works of Mark Twain*, rev. ed. (New York: Harper and Brothers, 1935), p.40.

A NOTE ON THE TEXT

Robert H. Hirst

This text of *The Prince and the Pauper* is a photographic facsimile of a copy of the first American edition dated 1882 on the title page. Although books printed from the first edition plates were manufactured until at least 1891, the earliest copies of the first edition were printed by November 15, 1881. Two copies were deposited with the Copyright Office on December 12, 1881, the official date of publication (despite the title page date of 1882). The copy reproduced here, which was inscribed on February 27, 1882, is an example of the second state, *Ab,* as defined by the editors of the California edition, for it contains three authorial corrections made for the first time in the second printing, which was completed on November 30, 1881: 'canopy of state' for the previous 'canopy of estate' (124); 'do think' for the erroneous 'do not think' (263); and 'reined' for the mistaken 'reigned' (362). This copy was therefore not among the first 10,030 copies off the press, but in all likelihood it was among the next 15,189 printed by December 24, 1881 (*The Prince and the Pauper,* ed. Victor Fischer and Lin Salamo, The Works of Mark Twain, University of California Press, 1979, pp. 401, 419–21). Jacob Blanck distinguished two binding states, A and B, but as this copy shows, they do not equate with the first and second states of the text (*BAL* 3402). The original volume is in the collection of the Mark Twain House in Hartford, Connecticut (810/C625pr/1882/c. 10).

THE MARK TWAIN HOUSE

The Mark Twain House is a museum and research center dedicated to the study of Mark Twain, his works, and his times. The museum is located in the nineteen-room mansion in Hartford, Connecticut, built for and lived in by Samuel L. Clemens, his wife, and their three children, from 1874 to 1891. The Picturesque Gothic-style residence, with interior design by the firm of Louis Comfort Tiffany and Associated Artists, is one of the premier examples of domestic Victorian architecture in America. Clemens wrote *Adventures of Huckleberry Finn*, *The Adventures of Tom Sawyer*, *A Connecticut Yankee in King Arthur's Court*, *The Prince and the Pauper*, and *Life on the Mississippi* while living in Hartford.

The Mark Twain House is open year-round. In addition to tours of the house, the educational programs of the Mark Twain House include symposia, lectures, and teacher training seminars that focus on the contemporary relevance of Twain's legacy. Past programs have featured discussions of literary censorship with playwright Arthur Miller and writer William Styron; of the power of language with journalist Clarence Page, comedian Dick Gregory, and writer Gloria Naylor; and of the challenges of teaching *Adventures of Huckleberry Finn* amidst charges of racism.

CONTRIBUTORS

Beverly R. David is professor emerita of humanities and theater at Western Michigan University in Kalamazoo. She is currently working on volume 2 of *Mark Twain and His Illustrators*, and on a Mark Twain mystery entitled *Murder at the Matterhorn*. She has written a number of sections on illustration for the *Mark Twain Encyclopedia* and her *Mark Twain and His Illustrators, Volume 1 (1869–1875)* was published in 1989. Dr. David resides in Allegan, Michigan, in the summer and Green Valley, Arizona, in the winter.

Everett Emerson, Alumni Distinguished Professor of English and American Studies, Emeritus, at the University of North Carolina, Chapel Hill, is the author of *The Authentic Mark Twain: A Literary Biography of Samuel L. Clemens* (1984), and the author or editor of seven other books. He is an honorary member of the Emily Dickinson Society of Japan and the American Studies Association of Thailand, has been named Honored Scholar by the Early American Literature Division of the Modern Language Association of America, and is the founder and an honored life member of the Mark Twain Circle of America. He lives in Chapel Hill.

Shelley Fisher Fishkin, professor of American Studies and English at the University of Texas at Austin, is the author of the award-winning books *Was Huck Black? Mark Twain and African-American Voices* (1993) and *From Fact to Fiction: Journalism and Imaginative Writing in America* (1985). Her most recent book is *Lighting Out for the Territory: Reflections on Mark Twain and American Culture* (1996). She holds a Ph.D. in American Studies from Yale University, has lectured on Mark Twain in Belgium, England, France, Israel, Italy, Mexico, the Netherlands, and Turkey, as well as throughout the United States, and is president-elect of the Mark Twain Circle of America.

Robert H. Hirst is the General Editor of the Mark Twain Project at The Bancroft Library, University of California at Berkeley. Apart from that, he has no other known eccentricities.

Judith Martin writes the syndicated column "Miss Manners" and is the author of two novels, *Gilbert* (1982) and *Style and Substance* (1986), as well as *Miss Manners' Guide to Excruciatingly Correct Behavior* (1982), *Miss Manners' Guide to Rearing Perfect Children* (1984), *Miss Manners' Guide for the Turn-of-the-Millennium* (1989), *Miss Manners on (Painfully Proper) Weddings* (1995), and *Miss Manners Rescues Civilization* (1996). A former reporter and critic for the *Washington Post*, Mrs. Martin lives in Washington, D.C.

Ray Sapirstein is a doctoral student in the American Civilization Program at the University of Texas at Austin. He curated the 1993 exhibition *Another Side of Huckleberry Finn: Mark Twain and Images of African Americans* at the Harry Ransom Humanities Research Center at the University of Texas at Austin. He is currently completing a dissertation on the photographic illustrations in several volumes of Paul Laurence Dunbar's poetry.

ACKNOWLEDGMENTS

There are a number of people without whom The Oxford Mark Twain would not have happened. I am indebted to Laura Brown, senior vice president and trade publisher, Oxford University Press, for suggesting that I edit an "Oxford Mark Twain," and for being so enthusiastic when I proposed that it take the present form. Her guidance and vision have informed the entire undertaking.

Crucial as well, from the earliest to the final stages, was the help of John Boyer, executive director of the Mark Twain House, who recognized the importance of the project and gave it his wholehearted support.

My father, Milton Fisher, believed in this project from the start and helped nurture it every step of the way, as did my stepmother, Carol Plaine Fisher. Their encouragement and support made it all possible. The memory of my mother, Renée B. Fisher, sustained me throughout.

I am enormously grateful to all the contributors to The Oxford Mark Twain for the effort they put into their essays, and for having been such fine, collegial collaborators. Each came through, just as I'd hoped, with fresh insights and lively prose. It was a privilege and a pleasure to work with them, and I value the friendships that we forged in the process.

In addition to writing his fine afterword, Louis J. Budd provided invaluable advice and support, even going so far as to read each of the essays for accuracy. All of us involved in this project are greatly in his debt. Both his knowledge of Mark Twain's work and his generosity as a colleague are legendary and unsurpassed.

Elizabeth Maguire's commitment to The Oxford Mark Twain during her time as senior editor at Oxford was exemplary. When the project proved to be more ambitious and complicated than any of us had expected, Liz helped make it not only manageable, but fun. Assistant editor Elda Rotor's wonderful help in coordinating all aspects of The Oxford Mark Twain, along with literature editor T. Susan Chang's enthusiastic involvement with the project in its final stages, helped bring it all to fruition.

I am extremely grateful to Joy Johannessen for her astute and sensitive copyediting, and for having been such a pleasure to work with. And I appreciate the conscientiousness and good humor with which Kathy Kuhtz Campbell heroically supervised all aspects of the set's production. Oxford president Edward Barry, vice president and editorial director Helen McInnis, marketing director Amy Roberts, publicity director Susan Rotermund, art director David Tran, trade editorial, design and production manager Adam Bohannon, trade advertising and promotion manager Woody Gilmartin, director of manufacturing Benjamin Lee, and the entire staff at Oxford were as supportive a team as any editor could desire.

The staff of the Mark Twain House provided superb assistance as well. I would like to thank Marianne Curling, curator, Debra Petke, education director, Beverly Zell, curator of photography, Britt Gustafson, assistant director of education, Beth Ann McPherson, assistant curator, and Pam Collins, administrative assistant, for all their generous help, and for allowing us to reproduce books and photographs from the Mark Twain House collection. One could not ask for more congenial or helpful partners in publishing.

G. Thomas Tanselle, vice president of the John Simon Guggenheim Memorial Foundation, and an expert on the history of the book, offered essential advice about how to create as responsible a facsimile edition as possible. I appreciate his very knowledgeable counsel.

I am deeply indebted to Robert H. Hirst, general editor of the Mark Twain Project at The Bancroft Library in Berkeley, for bringing his outstanding knowledge of Twain editions to bear on the selection of the books photographed for the facsimiles, for giving generous assistance all along the way, and for providing his meticulous notes on the text. The set is the richer for his advice. I would also like to express my gratitude to the Mark Twain Project, not only for making texts and photographs from their collection available to us, but also for nurturing Mark Twain studies with a steady infusion of matchless, important publications.

I would like to thank Jeffrey Kaimowitz, curator of the Watkinson Library at Trinity College, Hartford (where the Mark Twain House collection is kept), along with his colleagues Peter Knapp and Alesandra M. Schmidt, for having been instrumental in Robert Hirst's search for first editions that could be safely reproduced. Victor Fischer, Harriet Elinor Smith, and especially Kenneth M. Sanderson, associate editors with the Mark Twain Project, reviewed the note on the text in each volume with cheerful vigilance. Thanks are also due to Mark Twain Project associate editor Michael Frank and administrative assistant Brenda J. Bailey for their help at various stages.

I am grateful to Helen K. Copley for granting permission to publish photographs in the Mark Twain Collection of the James S. Copley Library in La Jolla, California, and to Carol Beales and Ron Vanderhye of the Copley Library for making my research trip to their institution so productive and enjoyable.

Several contributors — David Bradley, Louis J. Budd, Beverly R. David, Robert Hirst, Fred Kaplan, James S. Leonard, Toni Morrison, Lillian S. Robinson, Jeffrey Rubin-Dorsky, Ray Sapirstein, and David L. Smith — were particularly helpful in the early stages of the project, brainstorming about the cast of writers and scholars who could make it work. Others who participated in that process were John Boyer, James Cox, Robert Crunden, Joel Dinerstein, William Goetzmann, Calvin and Maria Johnson, Jim Magnuson, Arnold Rampersad, Siva Vaidhyanathan, Steve and Louise Weinberg, and Richard Yarborough.

Kevin Bochynski, famous among Twain scholars as an "angel" who is gifted at finding methods of making their research run more smoothly, was helpful in more ways than I can count. He did an outstanding job in his official capacity as production consultant to The Oxford Mark Twain, supervising the photography of the facsimiles. I am also grateful to him for having put me in touch via e-mail with Kent Rasmussen, author of the magisterial *Mark Twain A to Z*, who was tremendously helpful as the project proceeded, sharing insights on obscure illustrators and other points, and generously being "on call" for all sorts of unforeseen contingencies.

I am indebted to Siva Vaidhyanathan of the American Studies Program of the University of Texas at Austin for having been such a superb research assistant. It would be hard to imagine The Oxford Mark Twain without the benefit of his insights and energy. A fine scholar and writer in his own right, he was crucial to making this project happen.

Georgia Barnhill, the Andrew W. Mellon Curator of Graphic Arts at the American Antiquarian Society in Worcester, Massachusetts, Tom Staley, director of the Harry Ransom Humanities Research Center at the University of Texas at Austin, and Joan Grant, director of collection services at the Elmer Holmes Bobst Library of New York University, granted us access to their collections and assisted us in the reproduction of several volumes of The Oxford Mark Twain. I would also like to thank Kenneth Craven, Sally Leach, and Richard Oram of the Harry Ransom Humanities Research Center for their help in making HRC materials available, and Jay and John Crowley, of Jay's Publishers Services in Rockland, Massachusetts, for their efforts to photograph the books carefully and attentively.

I would like to express my gratitude for the grant I was awarded by the University Research Institute of the University of Texas at Austin to defray some of the costs of researching The Oxford Mark Twain. I am also grateful to American Studies director Robert Abzug and the University of Texas for the computer that facilitated my work on this project (and to UT systems analyst Steve Alemán, who tried his best to repair the damage when it crashed). Thanks also to American Studies administrative assistant Janice Bradley and graduate coordinator Melanie Livingston for their always generous and thoughtful help.

The Oxford Mark Twain would not have happened without the unstinting, wholehearted support of my husband, Jim Fishkin, who went way beyond the proverbial call of duty more times than I'm sure he cares to remember as he shared me unselfishly with that other man in my life, Mark Twain. I am also grateful to my family — to my sons Joey and Bobby, who cheered me on all along the way, as did Fannie Fishkin, David Fishkin, Gennie Gordon, Mildred Hope Witkin, and Leonard, Gillis, and Moss

Plaine — and to honorary family member Margaret Osborne, who did the same.

My greatest debt is to the man who set all this in motion. Only a figure as rich and complicated as Mark Twain could have sustained such energy and interest on the part of so many people for so long. Never boring, never dull, Mark Twain repays our attention again and again and again. It is a privilege to be able to honor his memory with The Oxford Mark Twain.

Shelley Fisher Fishkin
Austin, Texas
April 1996